REVIEW OF THE RESEARCH PROGRAM OF THE U.S. DRIVE PARTNERSHIP

Fifth Report

Committee on the Review of the Research Program of
the U.S. DRIVE Partnership, Phase 5

Board on Energy and Environmental Systems

Division on Engineering and Physical Sciences

A Consensus Study Report of

The National Academies of
SCIENCES · ENGINEERING · MEDICINE

THE NATIONAL ACADEMIES PRESS
Washington, DC
www.nap.edu

THE NATIONAL ACADEMIES PRESS 500 Fifth Street, NW Washington, DC 20001

This activity was supported by Grant No. DE-EE0002931 with the U.S. Department of Energy. Any opinions, findings, conclusions, or recommendations expressed in this publication do not necessarily reflect the views of any organization or agency that provided support for the project.

International Standard Book Number-13: 978-0-309-0-45687-6
International Standard Book Number-10: 0-309-45687-8
Digital Object Identifier: https://doi.org/10.17226/24717

Additional copies of this publication are available for sale from the National Academies Press, 500 Fifth Street NW, Keck 360, Washington, DC 20001; (800) 624-6242 or (202) 334-3313; http://www.nap.edu/.

Printed in the United States of America

Suggested citation: National Academies of Sciences, Engineering, and Medicine. 2017. *Review of the Research Program of the U.S. DRIVE Partnership: Fifth Report.* Washington, DC: The National Academies Press. doi: https://doi.org/10.17226/24717.

The National Academies of

SCIENCES • ENGINEERING • MEDICINE

The **National Academy of Sciences** was established in 1863 by an Act of Congress, signed by President Lincoln, as a private, nongovernmental institution to advise the nation on issues related to science and technology. Members are elected by their peers for outstanding contributions to research. Dr. Marcia McNutt is president.

The **National Academy of Engineering** was established in 1964 under the charter of the National Academy of Sciences to bring the practices of engineering to advising the nation. Members are elected by their peers for extraordinary contributions to engineering. Dr. C. D. Mote, Jr., is president.

The **National Academy of Medicine** (formerly the Institute of Medicine) was established in 1970 under the charter of the National Academy of Sciences to advise the nation on medical and health issues. Members are elected by their peers for distinguished contributions to medicine and health. Dr. Victor J. Dzau is president.

The three Academies work together as the **National Academies of Sciences, Engineering, and Medicine** to provide independent, objective analysis and advice to the nation and conduct other activities to solve complex problems and inform public policy decisions. The National Academies also encourage education and research, recognize outstanding contributions to knowledge, and increase public understanding in matters of science, engineering, and medicine.

Learn more about the National Academies of Sciences, Engineering, and Medicine at **www.nationalacademies.org**.

COMMITTEE ON THE REVIEW OF THE RESEARCH PROGRAM OF THE U.S. DRIVE PARTNERSHIP, PHASE 5

JOHN H. JOHNSON, Michigan Technological University, Houghton, *Chair*
ALEXIS T. BELL, NAS[1]/NAE,[2] University of California, Berkeley
DAVID BODDE, Clemson University, Clemson, South Carolina
NADY BOULES, NB Motors, LLC, Troy, Michigan
GLENN EISMAN, Eisman Technology Consultants, Bailey Island, Maine
DAVID E. FOSTER, University of Wisconsin, Madison
MATT FRONK, Matt Fronk & Associates, Honeoye Falls, New York
ROBERT NOWAK, Defense Advanced Research Projects Agency (retired), Rehoboth Beach, Delaware
VERNON P. ROAN, University of Florida, Gainesville[3]
BERNARD ROBERTSON, NAE, Daimler Chrysler Corporation (retired), Bloomfield Hills, Michigan
JAMES A. SPEAROT, Mountain Ridgeline Consulting, LLC, Breckenridge, Colorado
SATISH TAMHANKAR, Linde, LLC (retired), Scotch Plains, New Jersey
ALAN TAUB, NAE, University of Michigan, Ann Arbor
KATHLEEN C. TAYLOR, NAE, General Motors Corporation (retired), Fort Myers, Florida
BRIJESH VYAS, Bell Laboratories (retired), Warren, New Jersey

Staff

JAMES J. ZUCCHETTO, Senior Board/Program Director,[4] Board on Energy and Environmental Systems, Study Director
DANA CAINES, Financial Manager
LINDA CASOLA, Senior Program Assistant
LANITA JONES, Administrative Coordinator
E. JONATHAN YANGER, Research Associate

[1] NAS, National Academy of Sciences.
[2] NAE, National Academy of Engineering.
[3] Resigned April 11, 2016.
[4] Retired as Senior Board/Program Director in February 2017.

BOARD ON ENERGY AND ENVIRONMENTAL SYSTEMS

[1] NAE, National Academy of Engineering.
[2] NAS, National Academy of Sciences.
[3] Deceased on April 26, 2017.

Preface

This report contains the results of a review by the National Academies of Sciences, Engineering, and Medicine's Committee on Review of the Research Program of the U.S. DRIVE Partnership, Phase 5 (see Appendix A for biographical information on the committee members). The government/industry partnership known as U.S. DRIVE (*D*riving *R*esearch and *I*nnovation for *V*ehicle Efficiency and *E*nergy Sustainability) was formed in 2011. It is very much in line with the partnerships that preceded it—namely, the FreedomCAR and Fuel Partnership and, prior to that, the Partnership for a New Generation of Vehicles.

The U.S. DRIVE vision is that "American consumers have a broad range of affordable personal transportation choices that reduce petroleum consumption and significantly reduce harmful emissions from the transportation sector." Its mission is to "accelerate the development of precompetitive and innovative technologies to enable a full range of efficient and clean advanced light-duty vehicles, as well as related energy infrastructure." The Partnership is focused on advanced technologies for all light-duty passenger vehicles: cars, sport utility vehicles, crossover vehicles, pickups, and minivans. It also addresses technologies for hydrogen production, distribution, dispensing, and storage, and the interface and infrastructure issues associated with the electric utility industry for the support of battery electric vehicles and plug-in hybrid electric vehicles.

The National Academies Committee on Review of the Research Program of the U.S. DRIVE Partnership, Phase 5, reviewed the activities since the fourth review of the Partnership. The report provides an overview of the structure and management of the Partnership as well as the major achievements associated with the goals of the Partnership. Since the previous review Toyota, Hyundai, and Honda have made available within the United States a limited number of fuel cell vehicle sales or leases to the general public. General Motors, a U.S. DRIVE Partnership member, has reported plans for a 2020 rollout of its latest fuel cell vehicle. The development and deployment of roadworthy fuel cell vehicles is a major accomplishment and one that will help to identify remaining technical, cost, manufacturing, and infrastructure challenges. Though the cars are still in the late stages of development, the fact that the cars have advanced to this point is due in part to research and development coordination by the Partnership and its prior organizations, as well as from decades of funding of pertinent research projects by the Department of Energy (DOE) and Partnership members.

The committee appreciates the effort by the personnel from DOE, U.S. Council for Automotive Research, and all the companies and national laboratories that prepared presentations and hosted our visits. The help of these members of the Partnership enabled us to get the latest data and information, which was very important for the committee's preparation of this report.

John H. Johnson, *Chair*
Committee on Review of the Research
Program of the U.S. DRIVE Partnership, Phase 5

Acknowledgments

The Committee on the Review of the U.S. DRIVE Partnership, Phase 5, is grateful to the representatives of the U.S. DRIVE Partnership, including the Department of Energy (DOE), and to the representatives of the companies and national laboratories who contributed a significant amount of their time and effort to this National Academies of Sciences, Engineering, and Medicine study by giving presentations at meetings or responding to committee requests for information, as well as hosting members of the committee at site visits. The committee also acknowledges the valuable contributions of other individuals who provided information and presentations at the committee's open meetings. Appendix C lists all of those presentations.

The committee offers its special appreciation to Christy Cooper, Director, U.S. DRIVE Partnership, Office of Vehicle Technologies, DOE, for her significant contributions in coordinating responses to the questions and in making presentations to the committee. Finally, the chair wishes to recognize the committee members and the staff of the Board on Energy and Environmental Systems for organizing and planning the committee meetings, gathering information, and drafting sections of the report. Jim Zucchetto in particular has done an outstanding job of facilitating the work of the committee and helping it to write a focused and timely report. Linda Casola provided efficient and very helpful support to its meetings and the report production and LaNita Jones provided capable support to help finish the project and this report.

This Consensus Study Report was reviewed in draft form by individuals chosen for their diverse perspectives and technical expertise. The purpose of this independent review is to provide candid and critical comments that will assist the National Academies of Sciences, Engineering, and Medicine in making each published report as sound as possible and to ensure that it meets the institutional standards for quality, objectivity, evidence, and responsiveness to the study charge. The review comments and draft manuscript remain confidential to protect the integrity of the deliberative process.

We thank the following individuals for their review of this report:

Katherine Ayers, Proton OnSite,
Jay Baron, Center for Automotive Research,
R. Stephen Berry, NAS, University of Chicago,
Paul N. Blumberg, NAE, Ford Motor Co. (retired),
Dennis A. Corrigan, DC Energy Consulting, LLC,
John M. German, International Council for Clean Transportation,
Linos J. Jacovides, NAE, Michigan State University,
Trevor O. Jones, NAE, International Development Corporation,
JoAnn Milliken, Department of Energy (retired),
John S. Newman, NAE, University of California-Berkeley,
William F. Powers, NAE, Ford Motor Co. (retired), and
Mahdi Shahbakhti, Michigan Technological University.

Although the reviewers listed above provided many constructive comments and suggestions, they were not asked to endorse the conclusions or recommendations of this report nor did they see the final draft before its release. The review of this report was overseen by Douglas M. Chapin, NAE, MPR Associates, Inc., and Chris T. Hendrickson, NAE, Carnegie Mellon University. They were responsible for making certain that an independent examination of this report was carried out in accordance with the standards of the National Academies and that all review comments were carefully considered. Responsibility for the final content rests entirely with the authoring committee and the National Academies.

Contents

Figures and Tables

FIGURES

TABLES

Summary

The energy security, environmental, and economic issues associated with the transportation sector and with light-duty vehicles can be addressed in a number of ways. An important part of the nation's approach to reducing petroleum consumption and the environmental impact of light-duty vehicles is to improve automotive technology in a variety of ways that lead to higher fuel economy vehicles that are affordable. In addition, vehicles that can use alternative sources of energy, such as electricity or hydrogen, can have low greenhouse gas (GHG) and other emissions. Since the early 1990s the nation has formed government-industry partnerships to help accelerate the research and development (R&D) for light-duty vehicles (see Chapter 1).

This report by the National Academies of Sciences, Engineering, and Medicine (the Academies) Committee on the Review of the Research Program of the U.S. DRIVE Partnership, Phase 5 (the committee), presents the results of a review of the U.S. DRIVE (*D*riving *R*esearch and *I*nnovation for *V*ehicle Efficiency and *E*nergy Sustainability) Partnership, which was formed in 2011. U.S. DRIVE is very much in line with the partnerships that preceded it, namely, the FreedomCAR and Fuel Partnership and, prior to that, the Partnership for a New Generation of Vehicles (PNGV).[1] The PNGV focused on achieving a significant increase in fuel economy for a family sedan and resulted in unveiling three concept vehicles at the end of that program. Under President George W. Bush, a shift in the program took place toward addressing the challenges of developing technologies for hydrogen fuel as well as for fuel cell vehicle technologies. The FreedomCAR and Fuel

[1] The focus of the committee's review of technologies for light-duty vehicles is on the Department of Energy's (DOE's) research and development (R&D) programs that support the goals of U.S. DRIVE.

Partnership was established to address these challenges and to advance the technologies sufficiently so that a decision on the commercial viability of hydrogen fuel cell vehicles (HFCVs) could be made by 2015. As the Obama administration took office in early 2009, a redirection began to take place, with reduced R&D on hydrogen and fuel cell vehicles and increased attention directed toward technologies for the use of electricity to power light-duty vehicles, with emphasis on plug-in electric vehicles including plug-in hybrid electric vehicles (PHEVs) and all-electric vehicles (or battery electric vehicles [BEVs]). The Academies reviewed the PNGV seven times, from 1993 to 2001; the FreedomCAR and Fuel Partnership three times, between 2004 and 2010; and the U.S. DRIVE Partnership in 2011-2012. The U.S. DRIVE Partnership is considered a continuation of the FreedomCAR and Fuel Partnership and hence the current review a fifth (Phase 5) review.[2] The Partnership provides a forum to discuss precompetitive, technology-specific R&D needs; identify possible solutions; and evaluate progress toward jointly developed technical goals.[3] This process helps to inform the Department of Energy (DOE) on the precompetitive R&D that is carried out by DOE's Vehicle Technologies Office (VTO) and the Fuel Cell Technologies Office (FCTO). U.S. DRIVE and its member partners focus on precompetitive R&D that can help to accelerate the emergence of advanced technologies that are commercially feasible.

The U.S. DRIVE vision is that "American consumers have a broad range of affordable personal transportation choices that reduce petroleum consumption and significantly reduce harmful emissions from the transportation sector." Its mission is as follows: "Accelerate the development of pre-competitive and innovative technologies to enable a full range of efficient and clean advanced light-duty vehicles, as well as related energy infrastructure" (U.S. DRIVE, 2016).

The guidance for the work of the U.S. DRIVE Partnership as well as the priority setting and targets for needed research are provided by 12 joint industry/government technical teams, and working groups are formed as needed to address crosscutting issues. This structure has been demonstrated to be an effective means of identifying high-priority, long-term precompetitive research needs for each technology with which the Partnership is involved (see Chapters 1 and 2).

Technical areas in which R&D as well as technology validation programs have been pursued include the following:

- Internal combustion engines (ICEs) operating on conventional and various alternative fuels,
- Automotive fuel cell power systems,
- Hydrogen storage systems (especially onboard vehicles),

---

[2] See previous reports for background on the partnerships, the various technical areas, and issues that the partnerships have addressed (NRC, 2001, 2005, 2008, 2009, 2010, 2013). The background and introduction presented here derive and cite much from the previous Academies' review (NRC, 2013).

[3] The committee views precompetitive government R&D on technology as long-term, high-risk work with regard to its potential transition into commercial viability.

- Batteries and other forms of electrochemical energy storage,
- Electric propulsion systems,
- Hydrogen production and delivery, and
- Materials leading to vehicle weight reductions.

In each of these technology areas, specific research targets have been established, although some targets and time frames are undergoing revision. U.S. DRIVE oversight is provided by an Executive Steering Group (ESG), which is not a federal advisory committee as defined by the Federal Advisory Committee Act (FACA). It consists of the DOE's Assistant Secretary for Energy Efficiency and Renewable Energy (EERE) and a vice-presidential-level executive from each of the Partnership companies. The DOE EERE efforts are divided between the VTO and FCTO. The Partnership collaborates with other DOE offices within EERE and outside of EERE, as appropriate, and other agencies such as the U.S. Environmental Protection Agency, the U.S. Department of Defense, and the U.S. Department of Transportation on safety-related activities.

The U.S. DRIVE partners presently include three automotive companies, five energy companies, two electric power companies, and the Electric Power Research Institute, with the DOE providing the federal leadership.[4] Several associate members are also associated with the technical teams. The Partnership does not itself have a budget or conduct or fund R&D, but each partner makes its own decisions regarding the funding and management of its projects (see Chapters 1 and 2 for more detailed discussion of the organization of the Partnership).

This Summary provides overall comments and a brief discussion of the technical areas covered more completely in this report and presents the committee's main findings and recommendations.

OVERARCHING RECOMMENDATIONS

There are a number of issues that indicate to the committee that it is an opportune time for the Partnership to take stock of its strategic position and its focus. Significant technological advances are occurring in the private sector, for example, with the emergence of a variety of plug-in electric vehicles, with offerings of fuel cell vehicles, and with rapid advances in technology for autonomous vehicles. The U.S. petroleum import situation is changing rapidly, with the United States becoming much less dependent. In some cases, the technology targets are too near term for a precompetitive focus, and perhaps targets should be set for at least 2025, if not 2030, to develop those high-risk technologies that the private sector will not pursue. The Partnership is revisiting many of its technical targets. For hydrogen fuel cell vehicles, a critical barrier is the deployment of a hydrogen infrastructure, which is currently outside the precompetitive focus of the Partner-

[4] Note that Tesla was a partner until it withdrew from the Partnership in July 2016.

ship. Although the committee has avoided making budget recommendations, new targets and timelines and changes in emphasis or on new technologies may obviously affect the distribution of DOE funding, although that will have to be determined by DOE.

Progress and Barriers

Given that the Partnership exists primarily as a technical information exchange and serves as just one of the many inputs to DOE programs, where the budgets for all this activity reside, it is difficult to ascertain exactly which achievements are directly attributable to the Partnership. The Partnership points to the DOE FCTO/VTO Annual Merit Review for details on all its initiatives, where a wealth of valuable information can be found.

Nevertheless, significant progress has clearly been made since the National Research Council (NRC) Phase 4 review in the period 2011-2012 in many of the technical areas, including such areas as advanced combustion, hydrogen fuel cell durability and cost, and electric drive systems (motor, power electronics, and battery) and cost, the details of which can be found in Chapter 3. At the same time, market introduction of improved hybrid electric vehicles (HEVs) and BEVs, both by automotive manufacturers represented in U.S. DRIVE and others, indicates that much of this technology is migrating out of the precompetitive realm and into the marketplace. The HFCV, currently being introduced in limited numbers by foreign original equipment manufacturers (OEMs), and anticipated by 2020 by one U.S. OEM (GM[5]), is expected to follow a path to commercialization with its own unique challenges, including, for example, infrastructure development. Since the Partnership is exclusively dedicated to precompetitive R&D, it is important that, informed by the cradle-to-grave (C2G) studies, the portfolio of projects be regularly reviewed to ensure that the focus remains on precompetitive challenges and relevant technology enablers. The C2G studies are important to determine the full life-cycle impacts of different advanced vehicles and their fuel/energy sources (e.g., hydrogen, electricity, hydrocarbon fuels, etc.).

While some of the remaining challenges are purely technical, cost remains a formidable barrier for essentially all the technologies under development. The other notable barrier is the infrastructure challenge confronting hydrogen. Policy matters and deployment are by definition beyond the scope of the Partnership, but lack of infrastructure is arguably the biggest challenge to the widespread deployment of hydrogen fuel cell vehicles, and continued emphasis by DOE on infrastructure enablers as well as an implementation plan is vital, whether within the Partnership or not.

[5] See, for example, R. Truett, "Fuel Cell Puzzle Comes Together," *Automotive News*, October 11, 2016, http://www.autonews.com/article/20161011/BLOG06/310119999/fuel-cell-puzzle-comes-together.

Adequacy and Balance

During the past few years, the DOE EERE budget devoted to hydrogen and fuel cell-related activities, which include both stationary and automotive applications, has remained stable at around $100 million per year, while the budget devoted to non-hydrogen-related vehicle technology has gradually increased. For the fiscal year 2016, hydrogen and fuel cell-related work was $101 million per year and the VTO funding was $310 million per year.

Much of the VTO funding is for electric drive vehicles, especially batteries. Part of the VTO funding is for technologies for medium- and heavy-duty vehicles. Since fuel cell vehicles are inherently electric vehicles, much of the work on electric drivetrain and improved batteries is equally applicable to both plug-in electric vehicles and HFCVs. While the $100 million devoted to purely hydrogen and fuel cell technologies is a much smaller share of the total EERE budget, it is still felt by the committee to be appropriate as a share of the overall effort for projects supporting U.S. DRIVE targets and goals. Within that overall effort, priorities for funding may shift among technical areas as technical challenges change. Furthermore, there is hydrogen-related work in other parts of DOE outside of EERE, for example, the Office of Basic Energy Sciences, the Office of Fossil Energy, as well as the Advanced Research Projects Agency-Energy.

Management, Strategy, and Priority Setting

Prior reviews of the U.S. DRIVE Partnership have been critical of both the target setting process and the decision-making process, due to the lack of an overall total vehicle systems analysis approach. Prior reviews have found that systems analysis has been applied very effectively at the subsystem or micro level, but that overall total systems analysis guiding high-level Partnership direction was lacking. The Partnership has taken steps to improve both of these processes. In 2015 the Partnership established a target setting task force (TSTF) that operates in concert with the vehicle systems analysis technical team and the new C2G analytical working group. The committee applauds the creation of the C2G working group and the TSTF, as well as the adoption of a more robust target setting process based on total system analysis. These recent changes in approach appear to address the criticisms of previous reviews and are most welcome.

Overall, the Partnership has an increasingly robust consensus process for developing goals and targets and for providing guidance and input to DOE to inform the management of relevant DOE projects, and this process benefits greatly from the recent addition of overall strategic analysis. However, the supervision of those projects and the decisions made within them are a DOE EERE responsibility and not that of the Partnership.

The Partnership points to the types of decisions that *are* made by the Partnership as being, for example, focused within the portfolio, such as a decision (by the ESG) to emphasize work on low-carbon fuels.

Given these limitations on the scope of decision making within the Partnership, particularly those due to FACA, the committee feels that the processes applied within the technical teams, and overall at the ESG and the Joint Operations Group level, are robust and appropriate, and that the Partnership has successfully fulfilled its mandate for technical information exchange.

Finding S-1 (2-1 in Chapter 2). The committee finds that the response to prior recommendations regarding management of the Partnership, particularly the creation of the target setting task force and the cradle-to-grave (C2G) working group, and the adoption of a portfolio-based strategy, are welcome improvements, and the Partnership is well managed. However, the increased engagement by the Executive Steering Group has improved from unacceptable to barely adequate. Furthermore, the Partnership currently regards the C2G working group as only a "temporary, task-specific" group.

Finding S-2 (3-43 in Chapter 3). The cradle-to-grave life-cycle analysis model provides a major step forward in the ability of the U.S. DRIVE to advise the industry and the Department of Energy on program and policy choices. This model provides the capabilities and insights that will give the Partnership a useful management tool and, with further development, a strategic and policy capability.

Recommendation S-1 (2-1 in Chapter 2). The Executive Steering Group should meet more regularly than annually, perhaps at least quarterly, and participate directly in the portfolio analysis and target setting process for revised 2020 and new 2025 goals. Furthermore, the recently published cradle-to-grave study on vehicle-fuel pathways and follow-on work by the target setting task force and cradle-to-grave working group should be used proactively and specifically to help shape the overall Office of Energy Efficiency and Renewable Energy portfolio, and the cradle-to-grave working group should be transitioned from temporary to permanent status.

Recommendation S-2 (3-32 in Chapter 3). The cradle-to-grave model should be continually updated and, where possible, tailored to improve its ability to support senior policy makers. Resources appropriate to this task should be provided. This updating will be an ongoing project, and the Partnership should consider upgrading the ad hoc working group to a technical team.

Finding S-3 (2-2 in Chapter 2). Given the reality that the Partnership does not direct or manage DOE-funded programs, overlaps with other DOE programs, and has no budget, there remains considerable ambiguity over the precise scope of the Partnership and its relationship with other DOE activities.

Recommendation S-3 (2-2 in Chapter 2). The Partnership is urged to provide more transparency and clarity regarding those Department of Energy projects deemed wholly or partly within the U.S. DRIVE portfolio and the achievements truly attributable to the Partnership.

Strategic Issues Looking Forward

Three trends have emerged since the NRC Phase 4 review of the Partnership took place that could have strategic implications in the future:

1. The dramatic change in domestic energy production has rendered the Obama administration's objective of reducing oil imports by 50 percent almost moot. While criteria pollutants will always be a concern and require substantial technical development to mitigate, it seems likely that in the future greenhouse gas (GHG) goals will present the greatest challenge. With GHG emissions as a primary focus, the pathways (e.g., combinations of vehicles technologies and fuels) to achieve extremely aggressive goals are very limited and would suggest that Partnership-related projects be increasingly focused on those few pathways that offer a realistic chance of success in meeting those goals.

2. With numerous electric vehicles (HFCVs and BEVs) expected to enter the marketplace in the next few years, the consumer will be presented with a number of zero-emission vehicle (ZEV) options to select from. This transition can be expected to take many years, and infrastructure is among the greatest challenges in each case, but particularly with regard to hydrogen. Although deployment and infrastructure per se are beyond the scope of the Partnership, there remains a need for precompetitive work on technology enablers to reduce system cost, improve durability, and substantially lower the cost of delivered "green" hydrogen and electricity.

3. Although the precise impact on the U.S. DRIVE Partnership is unclear at this point, there is no doubt that the move toward connected and autonomous vehicles is dramatically accelerating. Somewhat related to this is the increasingly rapid proliferation of such personal mobility models as car sharing and ridesharing. While there does not appear to be an obvious connection between these trends and the current Partnership-related DOE portfolio, shared, autonomous, plug-in electric vehicles could contribute to the environmental and energy goals of U.S. DRIVE, and they deserve close scrutiny for their potential impact on the Partnership in the future.

Recommendation S-4 (4-1 in Chapter 4). The Executive Steering Group should identify appropriate changes in Partnership focus to reflect the impact of new personal mobility models, shrinking opportunities to achieve the aggressive greenhouse gas goals, the transition of many candidate technologies into the competitive domain, and the significant infrastructure challenges in providing hydrogen at fueling stations at a competitive cost—in particular, while retaining the focus on precompetitive technology enablers.

ADVANCED INTERNAL COMBUSTION ENGINES AND EMISSION CONTROLS

As noted in NRC (2013), advanced combustion and emission controls for ICEs are important because ICEs for transportation systems are going to be the dominant automotive technology for decades, whether in conventional vehicles, HEVs, PHEVs, or biofuel or natural gas vehicles. There is still much opportunity to reduce the fuel consumption and environmental impact of ICE-powered vehicles, so it is important to keep an active research program in this area. Developing the enhanced understanding and tools to do this pushes the state of the art in all engineering sciences.

Finding S-4 (3-2 in Chapter 3). The advanced combustion and emissions control technical team (ACECTT) has established stretch efficiencies goals for 2020 for peak and intermediate engine loads for the three types of engine power train systems they expect to be most prevalent in the near term: hybrid applications, naturally aspirated engine systems, and downsized boosted engine systems. The ACECTT is also engaged in research activities in chemical kinetic development and promoting a more fundamental understanding of the interaction between fuel characteristics—such as Research and Motor Octane number, heat of vaporization, and so on—and different engine operating conditions. This work is aimed at facilitating the integration of advanced kinetically controlled combustion processes, that is, low-temperature combustion, as part of the engine's operating map, which is considered a longer-term technology.

Finding S-5 (3-3 in Chapter 3). The ACECTT focus for both near- and longer-term research is centered on conventional four-stroke engine architectures. However, work on alternative engine architectures is taking place. Some of that work is under DOE funding, and claims are being made in the literature of potential efficiency and environmental impact improvements for these different engine architectures.

Recommendation S-5 (3-1 in Chapter 3). The advanced combustion and emissions control technical team should be proactive in seeking out and assessing data on the performance of alternative engine architectures that will allow benchmarking against those within their current research portfolio.

FUELS FOR INTERNAL COMBUSTION ENGINES

The committee was told that the VTO plans to "downselect" a specific spark ignition candidate fuel by 2017 and demonstrate an optimized kinetically controlled engine-fuel system by 2025. This is a very aggressive set of objectives. In the view of the committee, meeting the timing of the goals identified by DOE for advanced engine-fuel combinations, although well intended, will be difficult.

Although the current portfolio of projects will provide technical data to aid in making the selection, the process of choosing an optimized engine-fuel system will be a challenge. Each potential combination will have benefits and drawbacks. It is not too early in planning to identify the process and criteria for selecting an optimum system. In addition, outside the Partnership, DOE has established a Co-Optima initiative. Last, what are the plans for promoting the use of such an engine-fuel combination in commercial vehicles?

Finding S-6 (3-9 in Chapter 3). DOE has set an aggressive timeline for identifying an "optimized kinetically controlled" engine-fuel system. The Co-Optima program will presumably help in developing the data to establish such an optimized system, but DOE has not yet addressed how such a system would be implemented in the light-duty vehicle fleet.

Recommendation S-6 (3-2 in Chapter 3). The Department of Energy (DOE) should further explain how the Co-Optima program will lead to the introduction of an optimum engine-fuel system in commercial practice. The introduction of high-efficiency, low greenhouse gas (GHG) internal combustion engine technology into the marketplace may require fuel formulations that are different from today's commercial fuels. Engine manufacturers will not introduce vehicles that utilize advanced combustion systems without the assurance that suitable fuels are available for the new combustion technology. Reaching consensus between the DOE's Co-Optima program and U.S. DRIVE on the concept of an optimum engine and fuel is necessary, but not sufficient. A plan for introduction of advanced combustion systems and fuels designed to increase transportation energy efficiency and reduce carbon dioxide (CO_2) emissions is required.

FUEL CELLS

HFCVs have been in a development phase by the major automotive companies for decades. Their attractiveness when compared to conventional ICE technology is based on the direct conversion of chemical (hydrogen) to electrical energy via an electrochemical process and reduced environmental impact provided that the hydrogen is derived from "green" primary energy sources. The efforts to develop HFCVs by the major automotive companies have been significant, as is evident from the magnitude of the investments made by the individual automotive OEMs, the number of patents issued, and the engineering accomplishments to date. Notably, within this review period, a number of foreign OEMs (Toyota, Hyundai, and Honda) have either initiated membrane-based fuel cell vehicle sales or leases to the general public in the United States or have announced that vehicles will be available within the next few years. General Motors, Ford, and Fiat Chrysler, all three U.S. DRIVE Partnership members, do not currently have vehicle offerings, yet GM has been cited in the open literature as stating they are "on track" to produce their Gen 2 HFCV by 2020.[6] Recent activities by the aforementioned OEMs, foreign and domestic, demonstrate that HFCVs are in the late stages of development and are now ready for customer engagement, albeit at a modest level owing to limited production volume and hydrogen delivery and refueling infrastructure issues.

In addition to the infrastructure barrier, technical challenges remain before widespread market penetration and consumer acceptance of HFCVs are realized, but the current introduction of a limited number of HFCVs is encouraging. Such challenges have been outlined in prior Academies reviews (Phases 1 through 4) and though many have been resolved, meeting cost and fuel cell durability targets simultaneously remains the most critical barrier to overcome if HFCVs are to become viable, both technically and commercially. DOE is developing additional activities at the national laboratories with the creation of consortia that will help focus and coordinate the R&D.

Finding S-7 (3-16 in Chapter 3). Since the NRC Phase 4 review, Toyota, Hyundai, and Honda have made available within the United States a limited number of fuel cell vehicle sales or leases to the general public. U.S.-based OEMs, with significant input from the Partnership, although in different states of development, have advanced fuel cell technology to the point that at least one U.S. DRIVE Partnership OEM (General Motors) is anticipating a rollout of its fuel cell vehicle in 2020. The development and deployment of roadworthy fuel cell vehicles is a major accomplishment and one that will help to identify remaining technical, cost, manufacturing, and infrastructure challenges. Though the cars are still in the late stages of development, the fact that the cars have advanced to this point is due in part to R&D coordination by the Partnership and its prior organizations, as well as from decades of funding of pertinent research projects by the DOE and Partnership members.

Finding S-8 (3-17 in Chapter 3). With the U.S. OEMs in different states of fuel cell vehicle development, and with competitive dynamics emerging, selected Partnership (fuel cell) goals and targets are relevant to only some of the OEM members (e.g., platinum loadings). Furthermore, it appears that there is a fine line between what might be considered near- and long-term projects based on the state of development of a given OEM's technology.

Recommendation S-7 (3-5 in Chapter 3). The Partnership should evaluate projects for their near-term or long-term potential impact and assign technology readiness levels to them. The Partnership should continually assess its process for

[6] See, for example, B. Snavely, "GM, Honda to make hydrogen fuel cells at Michigan factory," *USA Today*, January 30, 2017, http://www.usatoday.com/story/money/cars/2017/01/30/general-motors-honda-fuel-cell-deal/97240096/ or R. Truett, "Fuel Cell Puzzle Comes Together," *Automotive News*, October 11, 2016.

prioritizing projects and should continue to address the longer-term, precompetitive (lower technology readiness level) objectives.

ONBOARD HYDROGEN STORAGE

The mission of the hydrogen storage technical team is to "accelerate research and innovation that will lead to commercially viable hydrogen-storage technologies that meet the U.S. DRIVE Partnership goals." Vehicle driving range and fueling time are important customer attributes for fuel cell vehicles. The objective is to achieve a driving range of at least 300 miles for a full range of light-duty vehicles and at the same time meet performance, packaging, cost, rapid fueling time, and safety requirements.

Finding S-9 (3-21 in Chapter 3). All the goals for onboard hydrogen storage have not been met, and basic scientific research has not produced an easy solution to date. Yet, onboard hydrogen storage is an issue for several technical teams and working groups beyond the hydrogen storage technical team, for example, the materials technical team, the fuel cells technical team, the hydrogen codes and standards technical team, and the hydrogen delivery technical team. As the technologies continue to mature, the need to merge activities can be expected to increase because vehicle performance parameters might be achieved through a wider range of options than gravimetric and volumetric hydrogen storage density alone.

Recommendation S-8 (3-9 in Chapter 3). The hydrogen storage technical team should increasingly work with the other technical teams even beyond those areas where overlap currently exists.

HYDROGEN

Regardless of the source of hydrogen, it is clear that for there to be the possibility of widespread penetration of HFCVs into the light-duty fleet, there must be the availability of hydrogen for refueling. Hydrogen production by natural gas reforming is currently a cost-effective option for near-term hydrogen requirements, and it also provides a pathway to reduced GHG emissions. To further reduce GHG emissions, the use of renewable sources of energy, such as biomass, wind, and solar, is required. Development of such technologies is the focus of the long-term R&D. However, delivery and dispensing of hydrogen is still prohibitively expensive and requires technological advances to meet the overall cost targets for the HFCV option to be viable in the future. Pressures of 700 bar for compressed hydrogen gas in onboard storage tanks is currently the accepted option for onboard storage. The delivery and dispensing of hydrogen needs to meet the corresponding requirements, that is, even higher pressure (e.g.,

875-900 bar) at the pump. Thus, the R&D focus has been to develop low-cost compression technologies and materials and concepts for high-pressure hydrogen storage and transport. There are several hurdles with this approach, and alternative new concepts need to be continuously developed.

While electricity infrastructure required for plug-in electric vehicles and related vehicle technology options already exist, hydrogen infrastructure is practically nonexistent. Therefore, market introduction of HFCVs faces a daunting challenge. Moreover, with ongoing improvements in engine and battery technologies, HFCVs will face increased competition, for example, with increasingly improved HEVs. The lack of hydrogen infrastructure can derail HFCV deployment. Some states, like California, Connecticut, and New York, along with companies like Toyota and Honda, are promoting infrastructure build-out by providing funding. But to date there are very few operating fueling stations to support the projected market for HFCVs. High station cost is an obvious barrier.

Finding S-10 (3-22 in Chapter 3). U.S. DRIVE does not have a cost target for dispensed hydrogen; it is instead considered within the scope of the U.S. DOE R&D program. The DOE cost target for dispensed hydrogen of less than \$4/kg H_2 is based on its calculation of threshold cost.[7] Since DOE calculated this cost in 2011, there have been changes in the base case values such as the fuel economy for hybrid electric vehicles.

Recommendation S-9 (3-10 in Chapter 3). The hydrogen threshold cost calculation, published by the Department of Energy in 2011, should be revised by taking into consideration the advances in competing hybrid vehicle technologies as well as any progress made with vehicular hydrogen fuel cells. This should be carefully assessed and addressed by the appropriate U.S. DRIVE teams as well as the Executive Steering Group to incorporate the implications in the Partnership plans.

Finding S-11 (3-29 in Chapter 3). Although industrial gas companies currently have the most experience with hydrogen production, delivery, and infrastructure, they are not core members of the Partnership; some of them serve as associate members of technical teams, but not on a consistent basis.

Recommendation S-10 (3-16 in Chapter 3). U.S. DRIVE should consider having industrial gas companies involved in hydrogen infrastructure activities as permanent members rather than as temporary associate members.

Recommendation S-11 (3-15 in Chapter 3). The Executive Steering Group should address issues (e.g., how will fueling stations be installed and by whom,

[7] Threshold cost is calculated so as to be competitive with other transportation options that are expected in 2020.

who will produce hydrogen, how will investments occur in fueling infrastructure without sufficient fuel cell vehicles on the road and vice versa, etc.) related to hydrogen infrastructure and assess U.S. DRIVE's role to formulate an action plan to address the issues and barriers.

ELECTRIC PROPULSION AND ELECTRICAL SYSTEMS

The electric drive (consisting of an electronic motor and an electronic controller) is a critical part of electrified power trains for light-duty vehicles. Therefore, a key objective of the U.S. DRIVE Partnership is the development of technologies addressing the electric drive component cost, weight, and size to help expedite electrified power train market penetration. Several motor configurations and design variations are under investigation to address the high cost of rare earth magnets. Significant progress was reported by the Partnership in power electronics. This was achieved by using innovative packaging and integration of classic inverters and converters and also by exploring the use of wide bandgap (WBG) devices for automotive power electronic systems. Also, pursuing WBG devices with the vigor and intensity shown in the DOE programs is commendable due to the size, weight, and efficiency benefits of using these devices in electrified vehicles. Given the inherent cost advantage of gallium nitride (GaN) devices grown on silicon (Si) substrates compared with silicon carbide (SiC) on SiC substrates (due to the much higher cost of SiC compared to Si), it is expected that GaN will ultimately be preferred among these two competing technologies for automotive applications. Historically, SiC devices have been the focus of research for many years, as they possess higher voltage and temperature capabilities than GaN devices. Operating at these high levels of temperatures requires other circuit components to be also capable of these temperatures, which is cost prohibitive for automotive applications but not so for other cost-tolerant applications, such as for defense.

Finding S-12 (3-33 in Chapter 3). Only a few projects are exploring GaN, with the majority focusing on SiC. Given GaN's potential cost advantage, it is expected to ultimately be the preferred choice for automotive applications.

Recommendation S-12 (3-20 in Chapter 3). The U.S. DRIVE Partnership should increase the focus on the advancement of gallium nitride technology in order to accelerate its readiness for commercial implementation.

ELECTROCHEMICAL ENERGY STORAGE

Improving electrochemical energy storage technologies, such as batteries, is needed for achieving the goals of the U.S. DRIVE Partnership. Batteries and supercapacitors are used in all electric drive vehicles including HEVs, PHEVs, BEVs, and HFCVs.

High cost remains the main impediment to significant market penetration of plug-in electric vehicles, which use large batteries. There is also a need to improve battery performance characteristics, that is, energy density, specific energy, operation at extreme temperatures, charging and discharging rates, cycle, and calendar life. These improvements in performance and cost reduction need to be realized while addressing the inherent safety issues associated with lithium batteries. Lithium-ion battery performance, cost, and safety are being addressed by DOE and other government entities, automotive OEMs, and battery manufacturers. DOE is exploring additional battery chemistries in order to surpass the performance and reduce the cost relative to lithium-ion batteries.

Finding S-13 (3-36 in Chapter 3). After a 20-year gap, a new set of energy storage goals and targets for the various electric vehicles was established in 2012. These targets are to be realized in the year 2020. This was a very positive step; however, all of the goals and targets are not in one place and there are several inconsistencies in the target values in various publications. It may be time to revisit the goals and targets given the advances in battery technology and vehicle implementation in the last 4 years.

Recommendation S-13 (3-23 in Chapter 3). U.S. DRIVE should establish a single, authoritative website for energy storage targets and goals for the various electric vehicle applications that is prominently and easily accessible to all. The dates that targets and goals were set or reviewed without change should be provided. The site should provide a roadmap of energy storage needs for several (rolling) decades into the future for use by research organizations and investigators for various applications and differing time frames.

GRID IMPACTS OF ELECTRICITY AS AN ENERGY SOURCE FOR VEHICLES

The convenience, affordability, and environmental impacts of electric energy have become an important consideration for the U.S. DRIVE Partnership. The environmental and energy security benefits from plug-in electric vehicles will increase in proportion to their use, commonly measured in electric vehicle miles traveled. And so, the availability and cost of recharging options weighs importantly in consumer decisions to purchase and use plug-in electric vehicles. Hydrogen fuel cell vehicles also rely on the electric grid, since the electrolysis of water could be used to produce hydrogen fuel. In addition, these vehicles could serve as a backup electric supply with a typical automotive power train, about 70 kW, able to serve a small cluster of homes.

Finding S-14 (3-38 in Chapter 3). The electric grid is beginning a period of disruptive change brought on by (1) technological opportunity, especially in microelectronics, deep learning, and robotics; (2) the global need to reduce the carbon emissions from electricity production; (3) marketplace demand for new and more efficient energy services; and (4) threats from outsiders to the security of a more automated grid linked to external devices. As a consequence, the electric grid will continue to evolve in ways that are not predictable to either the incumbents or the disruptors. State regulatory authorities will shape the pace and direction of this transition to a greater extent than the federal government.

Recommendations S-14 (3-25 in Chapter 3). The U.S. DRIVE partners should closely monitor the evolution of the electric grid to understand how (or whether) vehicle design can enable effective participation in the emerging electric marketplace in a way that increases the market share of nonpetroleum vehicles such as hydrogen fuel cell vehicles and (possibly) battery electric vehicles.

STRUCTURAL MATERIALS

As noted and discussed in Taub and Luo (2015), a major approach for improving vehicle efficiency, and thus fuel economy, is reducing the vehicle mass. A midsize family car weighs about 1,450 kg, and it takes a weight reduction of approximately 150 kg, or 10 percent, to achieve a 3 to 6 percent improvement in fuel economy. Following the global oil crisis of the 1970s, the weight of automobiles decreased consistently for about one decade. This was followed by a period of stable oil prices and, in the North American market, a shift to larger and heavier vehicles. Since the 1990s, engineering improvements in vehicle structural efficiency have continued, but the improvements have been offset by increased safety features and other consumer-driven content, such as convenience features and infotainment systems. More recently, higher fuel economy standards are being adopted worldwide, and the newest vehicle models are exhibiting weight reductions of 5 to 10 percent or more.

Finding S-15 (3-40 in Chapter 3). The U.S. DRIVE materials technical team (MTT), like the FreedomCAR and Fuel Partnership before it, has adopted a stretch goal of 50 percent reduction in vehicle weight (versus 2002 comparable vehicles) *with equal affordability* (emphasis added). Previous NRC (2010, 2013) committees found this goal to be unrealistic. The present committee agrees with that assessment. Further, during this review, the MTT reported a revision to the target by setting the comparator as a 2015 vehicle. The committee feels that this makes the target even more unrealistic.

Finding S-16 (3-41 in Chapter 3). The committee was pleased to see the MTT reported adoption of a midterm target for 2020 of an 18 percent weight glider reduction to be achieved at <$11/kg saved (<$5/pound) saved while maintaining equal vehicle-level performance (crash; noise, vibration, and harshness; durability; reliability; and recyclability).[8] However, the 2016 DOE Annual Merit Review referred to a 30 percent reduction by 2022 relative to a 2012 baseline. This is not realistic given the 2020 target. When designing a new vehicle, the options available for weight reduction are compared with other fuel economy improvement solutions including increased power train efficiency, vehicle electrification, decreased tire rolling resistance, and improved aerodynamics. It appears to be more appropriate for the Partnership to set a long-term target for weight reduction, in a similar manner.

Recommendation S-15 (3-29 in Chapter 3). U.S. DRIVE should set the long-term target for the cost of weight reduction to be consistent with the long-term cost targets for the other technical teams. The committee also recommends continuing the practice of setting midterm targets. In doing so, it is important for all Department of Energy and U.S. DRIVE sources to reference a consistent set of targets.

REFERENCES

NRC (National Research Council). 2001. *Review of the Research Program of the Partnership for a New Generation of Vehicles: Seventh Report*. Washington, D.C.: National Academy Press.

NRC. 2005. *Review of the Research Program of the FreedomCAR and Fuel Partnership: First Report*. Washington, D.C.: The National Academies Press.

NRC. 2008. *Review of the Research Program of the FreedomCAR and Fuel Partnership: Second Report*. Washington, D.C.: The National Academies Press.

NRC. 2009. *Letter Report on Review of the Research Program of the FreedomCAR and Fuel Partnership, Phase 3*. Washington, D.C.: The National Academies Press.

NRC. 2010. *Review of the Research Program of the FreedomCAR and Fuel Partnership: Third Report*. Washington, D.C.: The National Academies Press.

NRC. 2013. *Review of the Research Program of the U.S. DRIVE Partnership: Fourth Report*. Washington, D.C.: The National Academies Press.

Taub, A.I., and A. Luo. 2015. Developing Advanced Lightweight Materials and Processes—Applying Materials Science to Automotive Engineering. *MRS Bulletin* 40(12):1045-1054.

U.S. DRIVE. 2016. *US DRIVE, Driving Research and Innovation for Vehicle Efficiency and Energy Sustainability: Partnership Plan, November*. Washington, D.C.: U.S. Department of Energy.

[8] The glider is the vehicle structure excluding the power train. Historically, weight reductions are easier to achieve in that part of the vehicle.

1

Introduction

BACKGROUND

The U.S. government, through either the administration or the Congress, has generally addressed the supply of energy and its use because of national concerns related to energy independence, national security, the environment and sustainability, and affordability (NAS/NAE/NRC, 2009a). These goals are emphasized to one extent or another depending on the current Congress or administration.

The U.S. transportation sector and the use of light-duty vehicles (automobiles and light trucks) are almost completely dependent on petroleum as an energy source to power vehicles. Since the 1970s, petroleum imports satisfied part of this demand, at times reaching levels of 50 percent or more and in the view of policy makers represented an important U.S. national and energy security issue. However, in recent years, for example, 2010-2014, high global oil prices and the development of hydraulic fracturing ("fracking") helped to produce a boom in U.S. oil production from low-permeability geologic formations such as shale; the oil produced is referred to as "shale oil" or "tight oil." Recent projections by the Energy Information Administration (EIA) show a major change occurring in the U.S. dependence on energy imports (EIA, 2016). The combination of increased tight oil production and higher fuel efficiency for vehicles leads to EIA projecting declines in oil imports from 24 percent of demand in 2015 to 19 percent of demand in 2040 under EIA's reference case (EIA, 2016; Sieminski, 2016). Under EIA's high oil price scenario, the United States becomes a net exporter around 2025. So, the situation faced by the United States in recent decades has greatly changed. Another issue of concern is volatility. Although the price of petroleum, gasoline, and diesel dropped to low levels in late 2015 and early 2016, the economic environment since 2008 has been one of volatility. Significant economic impacts on the transportation sector, the automotive industry, the economy, and vehicle owners can arise from the price volatility of gasoline and diesel fuel.

In addition to these energy security and economic concerns, the automobile also has a significant environmental footprint as a consequence of tailpipe emissions. Furthermore, there are environmental impacts associated with the full life cycle of producing and delivering fuels to vehicles as well as the impacts of vehicle production and disposal if one looks at the sector from a full life-cycle perspective. The combustion of petroleum-derived fuels in the U.S. transportation sector, mostly gasoline and diesel, produces a significant fraction of the nation's anthropogenic greenhouse gases (GHGs), as well as such criteria pollutants as oxides of nitrogen (NO_x), nonmethane hydrocarbons, and particulate matter that affect local air quality (EPA, 2016a). Although criteria pollutant emissions from light-duty vehicles have declined dramatically in the past few decades because of improvements in engines, fuels, and emission control systems, there are still some areas of the country that are not in compliance with air quality standards. And as the number of vehicles increases, there continue to be concerns about emissions, especially in urban areas with high concentrations of vehicles. These concerns can be addressed with vehicles having zero tailpipe emissions, for example, with hydrogen fuel cell vehicles (HFCVs) or battery electric vehicles (BEVs).

In addition to concerns about criteria pollutants and their impact on local air quality is the desire to reduce GHG emissions that contribute to climate change. When the combustion of hydrocarbon fuels such as gasoline or diesel occurs in vehicle engines, carbon dioxide (CO_2) is produced, the major GHG contributing to global warming. The U.S. transportation sector accounted for about a third of total U.S. anthropogenic CO_2 emissions in 2014, and it is projected by the EIA to still constitute a significant fraction (35 percent) in 2040 (EPA, 2016b; EIA, 2016). Light-duty vehicles comprised about 60 percent of the CO_2 emissions from the transportation sector in 2014 (EPA, 2016b). If all GHGs are included (e.g., methane, nitrous oxide), the transportation sector accounted for about 26 percent of U.S. GHG emissions; therefore, the sector is an important source of GHG emissions, especially CO_2 emissions.

U.S. DRIVE PARTNERSHIP

The energy security, environmental, and economic issues associated with the transportation sector and with light-duty vehicles can be addressed in a number of ways. One particularly important approach, which is the subject of this report, is to improve light-duty vehicle technologies. For example, if engines are made more efficient and if the fuel economy of vehicles is improved, a vehicle's fuel consumption per mile can decline and the associated CO_2 emitted per mile will also decline. Hence, an important part of the nation's approach to reducing GHG emissions from light-duty vehicles is to improve automotive technology in a variety of ways that lead to higher fuel economy vehicles that are affordable. In addition, vehicles that can use alternative sources of energy, such as electricity or hydrogen, can have low GHG emissions if, for example, they are produced using renewable energy sources.

This report contains the results of a review by the National Academies of Sciences, Engineering, and Medicine's (the Academies) Committee on the Review of the Research Program of the U.S. DRIVE Partnership, Phase 5 (see Appendix A for biographical information on the committee members). The government/industry partnership known as U.S. DRIVE (*D*riving *R*esearch and *I*nnovation for *V*ehicle *E*fficiency and *E*nergy Sustainability) was formed in 2011. As noted in NRC (2013a), it is very much in line with the partnerships that preceded it, namely, the FreedomCAR and Fuel Partnership and, prior to that, the Partnership for a New Generation of Vehicles (PNGV). The Academies reviewed the PNGV seven times, from 1993 to 2001; the FreedomCAR and Fuel Partnership three times, between 2004 and 2010; and the U.S. DRIVE Partnership in 2011-2012. The U.S. DRIVE Partnership is considered a continuation of the FreedomCAR and Fuel Partnership and hence the current review a fifth (Phase 5) review. (See previous reports for background on the partnerships, the various technical areas, and issues that the partnerships have addressed [NRC, 2001, 2005, 2008a, 2009, 2010a,b, 2013a,b, 2015a,b].) The committee's report represents a continuing review of the partnerships that have been formed to address advanced light-duty vehicle and associated infrastructure challenges. The main charge to the committee for this report is to review activities since the fourth review of the U.S. DRIVE Partnership (NRC, 2013a). The full statement of task for the committee is provided later in this chapter.

As noted in NRC (2013a), for decades the Department of Energy (DOE) has funded and supported research and development (R&D) programs related to advanced vehicular technologies and alternative transportation fuels. Under the Clinton administration during the 1990s much of this R&D for light-duty vehicles was conducted under the PNGV. This initial government–auto industry partnership was formed between the federal government and the auto industry's U.S. Council for Automotive Research (USCAR).[1] The PNGV sought to improve the nation's competitiveness significantly in the manufacture of future generations of vehicles, to implement commercially viable innovations emanating from ongoing research on conventional vehicles, and to develop vehicles that achieve up to three times the fuel efficiency of comparable 1994 family sedans (DOE, 2004a,b,c; NRC, 2001; PNGV, 1995; The White House, 1993).

---

[1] USCAR, which predated PNGV, was established by Chrysler Corporation, Ford Motor Company, and General Motors Corporation. Its purpose was to support intercompany, precompetitive cooperation so as to reduce the cost of redundant R&D, especially in areas mandated by government regulation, and to make the U.S. industry more competitive with foreign companies. Chrysler Corporation merged with Daimler Benz in 1998 to form DaimlerChrysler. In 2007 DaimlerChrysler divested itself of a major interest in the Chrysler Group, and Chrysler LLC was formed, which became Chrysler Group LLC. Chrysler Group LLC then became FCA US LLC (Fiat Chrysler Automobiles).

The PNGV focused on achieving a significant increase in fuel economy for a family sedan and resulted in unveiling three concept vehicles at the end of that program. Under President George W. Bush a shift in the program took place toward addressing the challenges of developing hydrogen fuel technologies as well as fuel cell vehicle technologies. The FreedomCAR and Fuel Partnership[2] was established to address these challenges and to advance the technologies enough so that a decision on the commercial viability of hydrogen vehicles could be made by 2015. As the Obama administration took office in early 2009 a redirection began to take place, with reduced R&D on hydrogen and fuel cell vehicles and increased attention directed toward technologies for the use of electricity to power light-duty vehicles, with emphasis on plug-in electric vehicles, including plug-in hybrid electric vehicles (PHEVs), and all-electric vehicles (or BEVs). However, as budgets were appropriated by Congress, R&D continued across all technologies relevant to fuel cells and hydrogen, as well as those relevant to PHEVs and BEVs. In 2011, the FreedomCAR and Fuel Partnership morphed into the U.S. DRIVE Partnership, and a U.S. DRIVE Partnership Plan was formally released in February 2012 and updated in 2016 (U.S. DRIVE, 2016). Outside the Partnership, the federal interest in increasing the use of alternative fuels was exemplified by the creation by Congress of the Renewable Fuel Standard (RFS) in 2005 prescribing annual amounts of renewable fuels to be used in transportation. Furthermore, extensive R&D on the production of biofuels is undertaken in DOE's Bioenergy Technologies Office (BETO), which is in the Energy Efficiency and Renewable Energy (EERE) Office of Transportation and also outside the Partnership.

Building on participation in the previous partnerships, currently U.S. DRIVE includes the following partners:

- *Automobile industry:* U.S. Council for Automotive Research LLC (USCAR, the cooperative research organization for FCA US LLC, Ford Motor Company, and General Motors Company)[3];
- *Electric utility industry:* DTE Energy Company, Southern California Edison Company, and the Electric Power Research Institute;
- *Federal government:* U.S. Department of Energy; and

[2] In February 2003, before the announcement of the FreedomCAR and Fuel Partnership, President George W. Bush announced the FreedomCAR and Hydrogen Fuel Initiative to develop technologies for (1) fuel-efficient motor vehicles and light trucks, (2) cleaner fuels, (3) improved energy efficiency, and (4) hydrogen production, and a nationwide distribution infrastructure for vehicle and stationary power plants, to provide fuel for both hydrogen internal combustion engines and fuel cells (DOE, 2004b). The expansion of the FreedomCAR and Fuel Partnership to include the energy sector after the announcement of the initiative also supported the goal of the FreedomCAR and Hydrogen Fuel Initiative. The partners in the program included DOE, USCAR, BP America, Chevron Corporation, ConocoPhillips, ExxonMobil Corporation, and Shell Hydrogen (U.S.). During 2008, with increased interest in plug-in hybrid electric vehicles and battery electric vehicles, the electric utilities DTE Energy (Detroit) and Southern California Edison were added (DOE, 2009).

[3] Tesla Motors was a member but withdrew in July 2016.

- *Fuel industry:* BP America, Chevron Corporation, Phillips 66 Company, ExxonMobil Corporation, and Shell Oil Products U.S.

According to U.S. DRIVE (2016) and as noted in NRC (2013a), the Partnership is a nonbinding, nonlegal, voluntary government–industry partnership. It does not itself conduct or fund R&D, but each partner makes its own decisions regarding the funding and management of its projects. By bringing together technical experts and providing a framework for frequent and regular interaction, the Partnership provides a forum for discussing precompetitive, technology-specific R&D needs, identifies possible solutions, and evaluates progress toward jointly developed technical goals. Its frequent communication among partners also helps to identify potential duplication of efforts and increases the chances of successful commercialization of publicly funded R&D.[4] Most of the committee's review of technology development is focused on the DOE precompetitive R&D programs in the Vehicle Technologies Office (VTO) and in the Fuel Cell Technologies Office, both of which reside within the Office of Transportation, which is part of EERE (see Appendix B). See Chapter 2 for further discussion of the organization of the Partnership and how it functions.

The U.S. DRIVE (2016) vision is that

> American consumers have a broad range of affordable personal transportation choices that reduce petroleum consumption and significantly reduce harmful emissions from the transportation sector.

Its mission is to

> Accelerate the development of precompetitive and innovative technologies to enable a full range of efficient and clean advanced light-duty vehicles, as well as related energy infrastructure.

The Partnership is focused on advanced technologies for all light-duty passenger vehicles: cars, sport utility vehicles (SUVs), crossover vehicles, pickups, and minivans. It also addresses technologies for hydrogen production, distribution, dispensing, and storage, and the interface and infrastructure issues associated with the electric utility industry for the support of BEVs and PHEVs (NRC, 2013a). Furthermore, as noted in previous National Research Council (NRC) reviews, the activities and success of the Partnership "can serve as an inspiration and motivation for the next generation of scientists and engineers, and thus contribute to restoring American leadership in research and its application for the public good" (NRC, 2010a, p. 18, 2013a, p. 18).

The Partnership facilitates communication among its partners and examines precompetitive technologies in four broad categories, all of which include potential issues related to the technologies or fuels as follows (U.S. DRIVE, 2016):

---

[4] The committee views precompetitive government R&D on technology as long-term, high-risk work with regard to its potential transition into commercial viability.

- *Vehicles*
  - —Advanced combustion and emissions control,
  - —Fuel cells,
  - —Electrochemical energy storage (e.g., batteries),
  - —Electric drive and power electronics,
  - —Lightweight materials, and
  - —Vehicle systems and analysis.
- *Fuels*
  - —Hydrogen production,
  - —Hydrogen delivery,
  - —Fuel pathway integration, or
  - —Other sustainable mobility fuels as agreed to by the Partnership.
- *Joint vehicles/fuels*
  - —Hydrogen codes and standards, and
  - —Hydrogen storage.
- *Joint vehicles/electric utility*
  - —Electric grid interaction.

As also discussed in NRC (2013a), the Partnership addresses the technical challenges associated with the envisioned pathways by establishing quantitative performance and cost targets[5,6] for precompetitive technologies. These targets and the research related to their attainment are discussed later in this report. Technical teams, as discussed in Chapter 2, specify and manage technical and crosscutting needs of the Partnership. A technical team is associated with each of the bulleted areas noted earlier in the four broad categories. If special issues arise, working groups may be formed (see Chapter 2).

RECENT CHANGES SINCE THE PHASE 4 REVIEW

A number of changes in the regulatory environment, in automotive technology, and in the automotive marketplace have been occurring in recent years, some since the Academies issued its report in 2013 on the fourth review of the U.S. DRIVE Partnership. Industry has taken the lead in the development of fuel cell and plug-in electric vehicles (BEVs and PHEVs). As a result, a competitive commercial environment has arisen as BEVs, HFCVs, and PHEVs enter the marketplace from both domestic and foreign manufacturers, including HFCVs from foreign automotive companies. These changes are affecting the light-duty vehicle environment and indirectly may have some bearing on the strategy of the

[5] DOE defines "goals" as desired, qualitative results that collectively signify Partnership mission accomplishments. It defines "targets" as tangible, quantitative metrics to measure progress toward goals.

[6] All references to cost imply estimated variable cost (or investment, as appropriate) based on high volume (500,000 annual volume) unless otherwise stated. "Cost" refers to the cost of producing an item, whereas "price" refers to what the consumer would pay.

Partnership as it looks to future precompetitive R&D. Some of these changes are briefly reviewed in what follows.

The Regulatory Environment

As discussed in previous reviews by the National Academies, the U.S. government during the past few decades has enacted legislation and policies to help achieve its national goals in the transportation sector (NRC, 2013a). For example, the Corporate Average Fuel Economy (CAFE) regulations have increased and are projected to further increase the average miles per gallon (mpg) for light-duty vehicles and reduce GHG emissions, while federal emissions standards have led to a dramatic decrease in criteria vehicle emissions per mile traveled.[7] The increasing levels of CAFE standards have created a need for advanced automotive technologies that will increase the relevance of the precompetitive R&D directed at technology development in the U.S. DRIVE Partnership. Other legislation, as noted previously, such as the RFS, seeks to promote the replacement of petroleum-based fuels with alternative fuels, such as those derived from biomass (NRC, 2011b). Federal R&D helps enable advanced vehicle and fuel technologies to emerge in the commercial marketplace (NRC, 2011a), which can help to address the nation's energy security, economic, and environmental challenges. In fact, DOE developed a broad set of strategies in its Quadrennial Technology Reviews (QTRs) to address the nation's energy challenges, including electrifying the vehicle fleet and increasing vehicle efficiency (DOE, 2011, 2015). However, the challenges of doing so on a large scale are formidable.

As noted in NRC (2013a), in addition to the federal legislation noted above, California has programs to reduce emissions of GHGs from vehicles, one of which is the zero-emission vehicle (ZEV) program. The state is promoting the adoption of ZEVs—for example, electric vehicles and HFCVs—by setting benchmarks for 2020 and 2025 for infrastructure to support such vehicles as well as for the adoption of such vehicles. California's Executive Order B-16-2012 aims for there to

---

[7] In 2010 CAFE standards were enacted requiring light-duty vehicles (passenger cars and light trucks) to meet 35.5 mpg by model year (MY) 2016. In October 2012 the National Highway Traffic Safety Administration (NHTSA) and U.S. Environmental Protection Agency (EPA) issued a joint rule to further improve fuel economy and reduce greenhouse gas emissions. The first phase of NHTSA's rule is expected to require fuel economy levels of about 41 mpg in MY 2021. The second phase of the CAFE program, from 2022 to 2025, includes augural standards that are not final but are estimated to require about 49 mpg for MY 2025. EPA is projecting emission levels on an industry-wide average of 163 g/mile of CO_2 by 2025, which also includes consideration of air conditioning leakage and alternative refrigerants. EPA's estimates equate to 54.5 mpg if these emission levels were achieved solely through vehicle fuel efficiency. These standards represent about a doubling from pre-2010 standards that were 27.5 mpg (NHTSA, 2016a). On January 12, 2017, the EPA administrator signed a final determination to maintain the current GHG emission standards for MY 2022-2025 vehicles. See https://www.epa.gov/regulations-emissions-vehicles-and-engines/midterm-evaluation-light-duty-vehicle-greenhouse-gas-ghg#final-determination.

be 1.5 million ZEVs in California by 2025, with supporting infrastructure and a growing market (CARB, 2016). Nine states are following California and requiring automakers to produce zero-emission vehicles (C2ES, 2016). These programs are also stimulating the development of the advanced vehicle technologies that are under development by some of the partners in the U.S. DRIVE Partnership. Furthermore, California passed legislation in September 2016 to reduce GHG emissions by at least 40 percent below 1990 levels by 2030. The availability of low-cost natural gas and initiatives by states to promote renewable electric power technologies is leading to lower GHG emissions from the electric power sector. This will affect the full fuel cycle GHG emissions from plug-in electric vehicles.

The Obama administration also placed a strong emphasis on actions to address climate change and reduce U.S. GHG emissions, and the President's Climate Action Plan was issued by the administration in 2013 (The White House, 2013). In the Climate Action Plan then president Obama reiterated his 2009 commitment to reducing overall U.S. GHG emissions by 17 percent below 2005 levels by 2020. To achieve such a goal will require that light-duty vehicles achieve significant reductions in petroleum use and corresponding GHG emissions. As noted above, more stringent fuel economy standards for light-duty vehicles have been enacted, and stricter fuel consumption standards for medium- and heavy-duty trucks have also been promulgated. Very recently, at the 2015 United Nations Climate Change Conference, Conference of the Parties, Twenty-first session (COP-21), the United States and other countries reached a historic international agreement to holding the increase in the global average temperature to well below 2°C above preindustrial levels and to aim to reach global peaking of GHG emissions as soon as possible.[8] The intent of this agreement was for the United States to achieve an economy-wide target of reducing its GHG emissions by 26 to 28 percent below 2005 levels by 2025 and to make best efforts to reduce them by 28 percent (The White House, 2015) and reach 83 percent reductions by 2050. Other initiatives by the Obama administration included the development of electric supply technologies with reduced GHG emissions as well as incentives for their deployment. As pointed out in NRC (2013a), if a large-scale penetration of BEVs or PHEVs takes place, then the goal of reducing GHGs significantly will require an electricity production system that reduces such emissions significantly compared to the current U.S. electric power system.

Advanced Vehicles

In the past few years, the automotive marketplace has also seen a dramatic change in the diversity of new and advanced automotive technologies emerging, in large part stimulated by the increasing CAFE standards as well as the ZEV mandates, and the improvements that have occurred in batteries, motors, and power

[8] See, e.g., http://unfccc.int/paris_agreement/items/9485.php.

electronic components for use in vehicles. There are now many plug-in electric vehicle models (e.g., the Bolt, Leaf, plug-in Prius, Tesla, and Volt) being offered with substantial battery storage incorporating electrified power trains including pure electrics, that is, BEVs, as well as PHEVs (which also include an internal combustion engine [ICE]). The cost of these vehicles still remains high compared to their conventional ICE counterparts. In addition, a limited number of vehicles with electrified power trains using fuel cells and hydrogen stored on board are being made available. For example, Toyota has made the fuel cell vehicle Mirai available in California and plans on a production run of 3,000 in 2017, and it plans on a smaller fuel cell vehicle by 2019 in Japan, anticipating that it could be selling 30,000 vehicles per year globally by 2020 (Voelcker, 2016). Hyundai is offering the Tucson fuel cell vehicle for lease in California, and Honda is offering leases for the fuel cell vehicle Clarity. Toyota has announced its intention to have all its vehicles be carbon free by 2050.[9] One major impediment to a broad availability of fuel cell vehicles continues to be the lack of a hydrogen delivery and refueling infrastructure for providing fuel to these vehicles, as discussed later in Chapter 3.

As pointed out in NRC (2013a), it is likely that in the coming decades there will be a diversity of vehicles and fuels that are commercialized. Some options are lower risk and nearer term than others, and they all face different technical, cost, and market risks. These issues have been explored in depth in other reports and will not be repeated here (see, for example, NAS/NAE/NRC, 2009a,b; NRC, 2008a,b, 2009, 2010a,b, 2011a,b, 2013a,b, 2015a,b; NRC/NAE, 2004). These studies have concluded that, given the high-risk and uncertain nature of many of these technologies and the immense challenge of achieving deep reductions in GHGs and petroleum use, an R&D strategy pursuing a portfolio of possible technological options is the most prudent approach (NRC, 2013a).

Trends in Vehicle Automation and Smart Transportation

Technologies that have been developed independent of the Partnership and that are being pursued for reasons other than support of Partnership goals will nevertheless influence achievement of those goals. These emerging technologies include the following:

- Deep learning technology, a variant of artificial intelligence, allows inference from the accumulation of experience.
- Advanced computer chips, like NVIDIA's recently announced "Xavier," are beginning to close in on the standard of excellence set by Google: 50 trillion operations per second at under 10 watts of power.

---

[9] The website for the Toyota Environmental Challenge 2050 is http://www.toyota-global.com/sustainability/environment/challenge2050/, accessed October 20, 2016.

- Superior sensor technology on board the vehicle can relieve the computational burden by providing more precise data about immediate traffic conditions.

These technologies have advanced rapidly since the Academies' fourth review of the U.S. DRIVE Partnership and now have the capability to change the urban transportation system in ways that help realize the goals of the U.S. DRIVE Partnership. These advances have occurred largely outside the U.S. DRIVE Partnership.

This section summarizes the manner in which vehicle and systems technologies might change to achieve these vehicle advancements, especially within the urban transportation system. These changes can be summarized under the rubric *connected and autonomous vehicles* (CAVs), a term that encompasses a range of technological and infrastructure developments that will allow mainly self-operated vehicles to communicate with each other and with their surrounding environment. The term "connected" refers to vehicles acting in concert via computer/intelligence applications, while "autonomous" refers to a range of computerized functions that assist drivers with tasks like lane keeping and adaptive cruise control and that might eventually relieve human drivers from all operating tasks.

Figure 1-1 was developed by the Society of Automotive Engineers (SAE) to show how automation might develop in stages, ranging from a scale of 0 to 5 representing the level of automation, and illustrating how those levels might evolve. Levels 0 to 2 have the human driver monitoring the driving environment, while levels 3 to 5 involve an automated driving system that monitors the driving environment. The figure shows how the human driver and the system execute the various functions of the vehicle: steering and acceleration/deceleration; monitoring the driving environment; fallback performance of the dynamic driving task; and the system capability for various driving modes. Clearly, these levels will occur in stages, with level 1 already occurring in new vehicles.

Five implications seem most important for U.S. DRIVE:

1. With the aggregation of human populations in urban areas, especially the large mega-cities, optimized service from shared, autonomous, plug-in electric vehicles (possibly hydrogen too) could do much to achieve the environmental and energy goals of the U.S. DRIVE Partnership.
2. Most discernible pathways through the transition will require the active participation of metropolitan transportation authorities, who can also be considered to be customers for automated mobility. Solutions are likely to reflect the local economic, demographic, and cultural characteristics of each jurisdiction.
3. The safety of vehicle occupants and bystanders has become a primary concern of regulators, and the NHTSA has recently released guidelines that provide a general framework for future safety requirements (NHTSA,

SAE Level	Name	Narrative Definition	Execution of Steering and Acceleration/Deceleration	Monitoring of Driving Environment	Fallback Performance of Dynamic Driving Task	System Capability (Driving Modes)
Human driver monitors the driving environment						
0	**No Automation**	The full-time performance by the human driver of all aspects of the dynamic driving task, even when enhanced by warning or intervention systems	Human driver	Human driver	Human driver	n/a
1	**Driver Assistance**	The driving mode-specific execution by a driver assistance system of either steering or acceleration/deceleration using information about the driving environment and with the expectation that the human driver perform all remaining aspects of the dynamic driving task	Human driver and system	Human driver	Human driver	Some driving modes
2	**Partial Automation**	The driving mode-specific execution by one or more driver assistance systems of both steering and acceleration/deceleration using information about the driving environment and with the expectation that the human driver perform all remaining aspects of the dynamic driving task	**System**	Human driver	Human driver	Some driving modes
Automated driving system ("system") monitors the driving environment						
3	**Conditional Automation**	The driving mode-specific performance by an automated driving system of all aspects of the dynamic driving task with the expectation that the human driver will respond appropriately to a request to intervene	System	**System**	Human driver	Some driving modes
4	**High Automation**	The driving mode-specific performance by an automated driving system of all aspects of the dynamic driving task, even if a human driver does not respond appropriately to a request to intervene	System	System	**System**	Some driving modes
5	**Full Automation**	The full-time performance by an automated driving system of all aspects of the dynamic driving task under all roadway and environmental conditions that can be managed by a human driver	System	System	System	**All driving modes**

FIGURE 1-1 Levels of automation for on-road vehicles.
SOURCE: SAE International, Standard J3016: Taxonomy and Definitions for Terms Related to On-Road Motor Vehicle Automated Driving Systems, http://www.sae.org/autodrive, 2014. Copyright © SAE International (2014).

2016b). These standards will influence the markets in which level 4 and 5 automated vehicles first deploy and set the pace of that deployment.

4. A second issue for the fully automated vehicles (levels 4 and 5 in the SAE levels of automation in Figure 1-1) concerns the transition, when the CAVs must share the limited roadway space with human drivers. The automated vehicles (always rational, attentive, and unemotional) must compete for right of way with human-driven vehicles (sometimes rational, frequently inattentive, and often aggressive). In many traffic situations, humans and automated systems must make joint decisions under levels of uncertainty that cannot be programmed in advance. The issue is less safety than the level of services that a fully automated vehicle can provide.[10]
5. Emerging business models for innovation are being built around ad hoc organizations termed "innovation ecosystems." These ecosystems can serve well in markets where technological advances occur rapidly and unpredictably and where customer demand is highly uncertain (Williamson and De Meyer, 2012). Many innovation ecosystems are being built through the acquisition of startup companies by industry incumbents: for example, Ford has invested $182 million in Pivotal Software, a cloud-computing venture; and Google has acquired four startup companies with the deep learning technologies since 2013, namely, DeepMind, Vision Factory, Dark Blue Labs, and DNNresearch. These new innovation models can move technology into the marketplace more rapidly than the traditional R&D model and so are relevant to the members of the Partnership.

FUNDING

The U.S. DRIVE Partnership is not funded as a line item in the federal budget. As discussed in Chapter 2 and noted in the current chapter, it is a means for exchanging information among the partners, eliciting various opinions on R&D directions, and helping to identify potential duplicative efforts and set targets for DOE technology development. Thus, it does not have a budget. The precompetitive R&D is under the control of DOE and, as noted in this chapter, the two main DOE offices that conduct technology R&D for light-duty vehicles are the VTO and the Fuel Cell Technologies Office (FCTO). The Vehicle Technologies Office was funded at a level of about $280 million in fiscal year (FY) 2015 and $310 million in FY 2016. The VTO pursues R&D not only for light-duty vehicle technologies but also for medium- and heavy-duty vehicles. The Fuel Cell

[10] Consider a CAV entering New York's Holland Tunnel, for example. To enter the city through this tunnel, motorists first queue up in eight lanes for the tollbooths. After paying, the traveled way reduces quickly to two lanes. The rules by which human drivers assign themselves priority reflect individual behaviors and hence are ambiguous. Humans are adept at navigating such ambiguities; robots are not. Fully driverless cars could be at a serious disadvantage in competition with a majority of human-driven vehicles.

Technologies Office was funded at a level of about $97 million in FY 2015 and $101 million in FY 2016. In reviewing the efforts and projects in these offices associated with the Partnership, the committee reviewed projects that the Partnership defined as associated with helping to meet its goals. The DOE budgets of various R&D activities within these two offices will be presented in the various sections in Chapter 3 that discuss the technologies.

COMMITTEE APPROACH AND ORGANIZATION OF THIS REPORT

The statement of task for this committee is as follows:

1. Review the challenging high-level technical goals and timetables for government and industry R&D efforts, which address such areas as (a) integrated systems analysis; (b) fuel cell power systems; (c) hydrogen storage systems; (d) hydrogen production and distribution technologies necessary for the viability of hydrogen-fueled vehicles; (e) the technical basis for codes and standards; (f) electric propulsion systems; (g) lightweight materials; (h) electric energy storage systems; (i) vehicle-to-grid interaction; and (j) advanced combustion and emission control systems for internal combustion engines.
2. Review and evaluate progress and program directions since the Phase 4 review toward meeting the Partnership's technical goals, and examine ongoing research activities and their relevance to meeting the goals of the Partnership.
3. Examine and comment on the overall balance and adequacy of the research and development effort, and the rate of progress, in light of the technical objectives and schedules for each of the major technology areas.
4. Examine and comment, as necessary, on the appropriate role for federal involvement in the various technical areas under development, especially in light of activities ongoing in the private sector or in the states.
5. Examine and comment on the Partnership's strategy for accomplishing its goals, especially in the context of ongoing developments across the portfolio of advanced vehicle technologies (e.g., biofuels, plug-in hybrid electric vehicles, electric vehicles), the recent enactment of legislation on corporate average fuel economy standards for light-duty vehicles, and possible legislation on carbon emissions. Other issues that the committee might address include: (a) program management and organization; (b) the process for setting milestones, research directions, and making Go/No Go decisions; (c) collaborative activities needed to meet the Partnership's goals (e.g., among the various offices and programs in DOE, the U.S. Department of Transportation, USCAR, the fuels industry, electric power sector, universities, and other parts of the private sector [such as venture capitalists], and others); and (d) other topics that the committee finds important to comment on related to the success of the Partnership to meet its technical goals.
6. Review and assess the actions that have been taken in response to recommendations from the Phase 4 review of the U.S. DRIVE Partnership.
7. Write a report documenting its findings and recommendations.

The committee met four times in face-to-face meetings to hear presentations from DOE and industry representatives involved in the Partnership and to discuss insights gained from the presentations and the written material gathered by the committee, and to work on drafts of its report (see Appendix C for a list of committee meetings and presentations). The committee established subgroups to investigate specific technical areas and formulate questions for DOE and other U.S. DRIVE partners to answer.

The committee subgroups also held several conference calls and site visits to collect information on technology development and other program issues. Some members of the committee also attended DOE's Annual Merit Review (AMR) or served as AMR reviewers in June 2016. Although the committee organized itself into subgroups, the entire committee participated in the final report and the findings and recommendations were agreed to by the whole committee. The Partnership also provided responses to the recommendations from the NRC Phase 4 report, and these are included in the National Academies public access file. DOE budget information included in this report was collected from presentations made to the committee (see Appendix C) as well as from information provided by the Partnership to committee questions. The information gathered enabled the committee to compose and reach consensus on this report.

The Summary presents the committee's main findings and recommendations. This chapter (Chapter 1) provides background on the Partnership and on its organization. Chapter 2 examines the management of the Partnership and the decision-making processes. Chapter 3 looks more closely at R&D for the various vehicle and fuel technologies that are of interest to the Partnership. Last, Chapter 4 presents an overall assessment of the Partnership's efforts and comments on some key issues. Appendix D contains a list of acronyms.

REFERENCES

CARB (California Air Resources Board). 2016. "Zero Emission Vehicle (ZEV) Program." July 15. https://www.arb.ca.gov/msprog/zevprog/zevprog.htm. Accessed September 12, 2016.

C2ES (Center for Climate and Energy Solutions). 2016. "Transportation Sector: ZEV Program." http://www.c2es.org/us-states-regions/policy-maps/zev-program. Accessed September 12, 2016.

DOE (U.S. Department of Energy). 2004a. *FreedomCAR and Vehicle Technologies Multi-Year Program Plan.* Washington, D.C.: U.S. Department of Energy, Energy Efficiency and Renewable Energy. http://www1.eere.energy.gov/vehiclesandfuels/resources/fcvt_mypp.html.

DOE. 2004b. *Hydrogen, Fuel Cells and Infrastructure: Multi-Year Research, Development and Demonstration Plan.* DOE/GO-102003-1741. Washington, D.C.: U.S. Department of Energy, Energy Efficiency and Renewable Energy. http://www.eere.energy.gov/hydrogenandfuelcells/mypp/.

DOE. 2004c. *Partnership Plan: FreedomCAR and Fuel Partnership.* Washington, D.C.: U.S. Department of Energy, Energy Efficiency and Renewable Energy. http://www.eere.energy.gov/vehiclesandfuels/pdfs/program/fc_fuel_partnership_plan.pdf.

DOE. 2009. "Addendum to the FreedomCAR and Fuel Partnership Plan to Integrate Electric Utility Industry Representatives." Washington, D.C.: U.S. Department of Energy, Energy Efficiency and Renewable Energy. February. http://www1.eere.energy.gov/vehiclesandfuels/pdfs/program/fc_fuel_addendum_2-09.pdf.

DOE. 2011. *Report on the First Quadrennial Technology Review.* DOE/S-0001. Washington, D.C. September 27. http://energy.gov/sites/prod/files/QTR_report.pdf.

DOE. 2015. *Quadrennial Technology Review: An Assessment of Energy Technologies and Research Opportunities.* Washington, D.C. September 15. http://www.energy.gov/sites/prod/files/2015/09/f26/Quadrennial-Technology-Review-2015_0.pdf.

EIA (U.S. Energy Information Administration). 2016. *Annual Energy Outlook 2016*. Washington, D.C. http://www.eia.gov/forecasts/aeo/. Accessed August 19, 2016.

EPA (U.S. Environmental Protection Agency). 2016a. "Air Pollutant Emissions Trends Data." June 29. https://www.epa.gov/air-emissions-inventories/air-pollutant-emissions-trends-data. Accessed September 12, 2016.

EPA. 2016b. *Inventory of U.S. Greenhouse Gas Emissions and Sinks: 1990-2014*. EPA 430-R-16-002. Washington, D.C. April 15. https://www3.epa.gov/climatechange/ghgemissions/sources.html. Accessed August 19, 2016.

NAS/NAE/NRC (National Academy of Sciences/National Academy of Engineering/National Research Council). 2009a. *America's Energy Future: Technology and Transformation*. Washington, D.C.: The National Academies Press.

NAS/NAE/NRC. 2009b. *Liquid Transportation Fuels from Coal and Biomass: Technological Status, Costs, and Environmental Impacts*. Washington, D.C.: The National Academies Press.

NHTSA (National Highway Traffic Safety Administration). 2016a. "CAFE-Fuel Economy." http://www.nhtsa.gov/Laws-&-Regulations/CAFE-%E2%80%93-Fuel-Economy. Accessed October 20, 2016.

NHTSA. 2016b. *Federal Automated Vehicles Policy: Accelerating the Next Revolution in Roadway Safety*. http://www.nhtsa.gov/nhtsa/av/index.html. Accessed October 20, 2016.

NRC (National Research Council). 2001. *Review of the Research Program of the Partnership for a New Generation of Vehicles: Seventh Report*. Washington, D.C.: National Academy Press.

NRC. 2005. *Review of the Research Program of the FreedomCAR and Fuel Partnership: First Report*. Washington, D.C.: The National Academies Press.

NRC. 2008a. *Review of the Research Program of the FreedomCAR and Fuel Partnership: Second Report*. Washington, D.C.: The National Academies Press.

NRC. 2008b. *Transitions to Alternative Transportation Technologies—A Focus on Hydrogen*. Washington, D.C.: The National Academies Press.

NRC. 2009. *Letter Report on Review of the Research Program of the FreedomCAR and Fuel Partnership, Phase 3*. Washington, D.C.: The National Academies Press.

NRC. 2010a. *Review of the Research Program of the FreedomCAR and Fuel Partnership: Third Report*. Washington, D.C.: The National Academies Press.

NRC. 2010b. *Transitions to Alternative Transportation Technologies—Plug-in Hybrid Electric Vehicles*. Washington, D.C.: The National Academies Press.

NRC. 2011a. *Assessment of Fuel Economy Technologies for Light-Duty Vehicles*. Washington, D.C.: The National Academies Press.

NRC. 2011b. *Renewable Fuel Standard: Potential Economic and Environmental Effects of U.S. Biofuel Policy*. Washington, D.C.: The National Academies Press.

NRC. 2013a. *Review of the Research Program of the U.S. DRIVE Partnership, Fourth Report*. Washington, D.C.: The National Academies Press.

NRC. 2013b. *Transitions to Alternative Vehicles and Fuels*. Washington, D.C.: The National Academies Press.

NRC. 2015a. *Cost, Effectiveness, and Deployment of Fuel Economy Technologies for Light-Duty Vehicles*. Washington, D.C.: The National Academies Press.

NRC. 2015b. *Overcoming Barriers to Deployment of Plug-in Electric Vehicles*. Washington, D.C.: The National Academies Press.

NRC/NAE (National Research Council/National Academy of Engineering). 2004. *The Hydrogen Economy: Opportunities, Costs, Barriers, and R&D Needs*. Washington, D.C.: The National Academies Press.

PNGV (Partnership for a New Generation of Vehicles). 1995. *Partnership for a New Generation of Vehicles Program Plan*. Draft. Washington, D.C.: U.S. Department of Commerce, PNGV Secretariat.

SAE International (Society of Automotive Engineers). 2014. "Automated Driving: Levels of Driving Automation Are Defined in New SAE International Standard J3016." http://www.sae.org/misc/pdfs/automated_driving.pdf.

Sieminski, A. 2016. "Annual Energy Outlook 2016: AEO 2016 Rollout." Paper presented at Johns Hopkins School of Advanced International Studies, Washington, D.C., June 28. http://www.eia.gov/pressroom/presentations/sieminski_06282016.pdf.

U.S. DRIVE. 2016. *US DRIVE, Driving Research and Innovation for Vehicle Efficiency and Energy Sustainability: Partnership Plan.* Washington, D.C.: U.S. Department of Energy. http://energy.gov/sites/prod/files/2016/08/f33/U%20S%20%20DRIVE%20Partnership%20Plan%20-%20July%202016.pdf.

Voelcker, J. 2016. Smaller, cheaper Toyota Mirai fuel-cell car coming in 2019, company says. *GreenCar Reports*, May 9. http://www.greencarreports.com/news/1103847_smaller-cheaper-toyota-mirai-fuel-cell-car-coming-in-2019-company-says.

The White House. 1993. *Historic Partnership Forged with Automakers Aims for Threefold Increase in Fuel Efficiency in as Soon as Ten Years.* Washington, D.C.: The White House.

The White House. 2013. "The President's Climate Action Plan." Washington, D.C. June 25. https://www.whitehouse.gov/share/climate-action-plan.

The White House. 2015. "Fact Sheet: U.S. Reports its 2025 Emissions Target to the UNFCCC." Washington, D.C. March 31. https://www.whitehouse.gov/the-press-office/2015/03/31/fact-sheet-us-reports-its-2025-emissions-target-unfccc.

Williamson, P.J., and A. De Meyer. 2012. Ecosystem advantage: How to successfully harness the power of providers. *California Management Review* 55(1):24-46.

2

Management, Strategy, and Priority Setting

ORGANIZATION OF THE PARTNERSHIP[1]

The U.S. DRIVE (*D*riving *R*esearch and *I*nnovation for *V*ehicle Efficiency and *E*nergy Sustainability) Partnership consists of a number of oversight groups and technical teams that have participants from government and industry. The Executive Steering Group (ESG), which is not a federal advisory committee as defined by the Federal Advisory Committee Act, is responsible for the governance of the Partnership and is made up of the Department of Energy (DOE) Assistant Secretary for the Office of Energy Efficiency and Renewable Energy (EERE) and a vice-presidential-level executive for each of the partnership companies. The ESG meets annually, and its defined role is to set high-level technical and management priorities for the Partnership.

Each of the three industry-related operations groups—the Vehicle Operations Group, the Fuel Operations Group, and the Electric Utility Operations Group—meets regularly on a schedule to suit the group's own needs. The Joint Operations Group (JOG) meets on a monthly basis to support the ESG, provide direction to the technical teams, and ensure strong coordination and a common understanding across the Partnership.

This structure, shown in Figure 2-1, is very much the same as existed in the predecessor FreedomCAR and Fuel Partnership and indeed, as existed in the U.S. DRIVE Partnership in 2013 when last reviewed by the National Research Council (NRC). Comparing Figure 2-1 with Figure 1-2 in the prior committee report (NRC, 2013), it is clear that the only significant change in the organization (other than

[1] Much of this section is taken from the NRC Phase 4 report (NRC, 2013) to succinctly describe the organization of the Partnership.

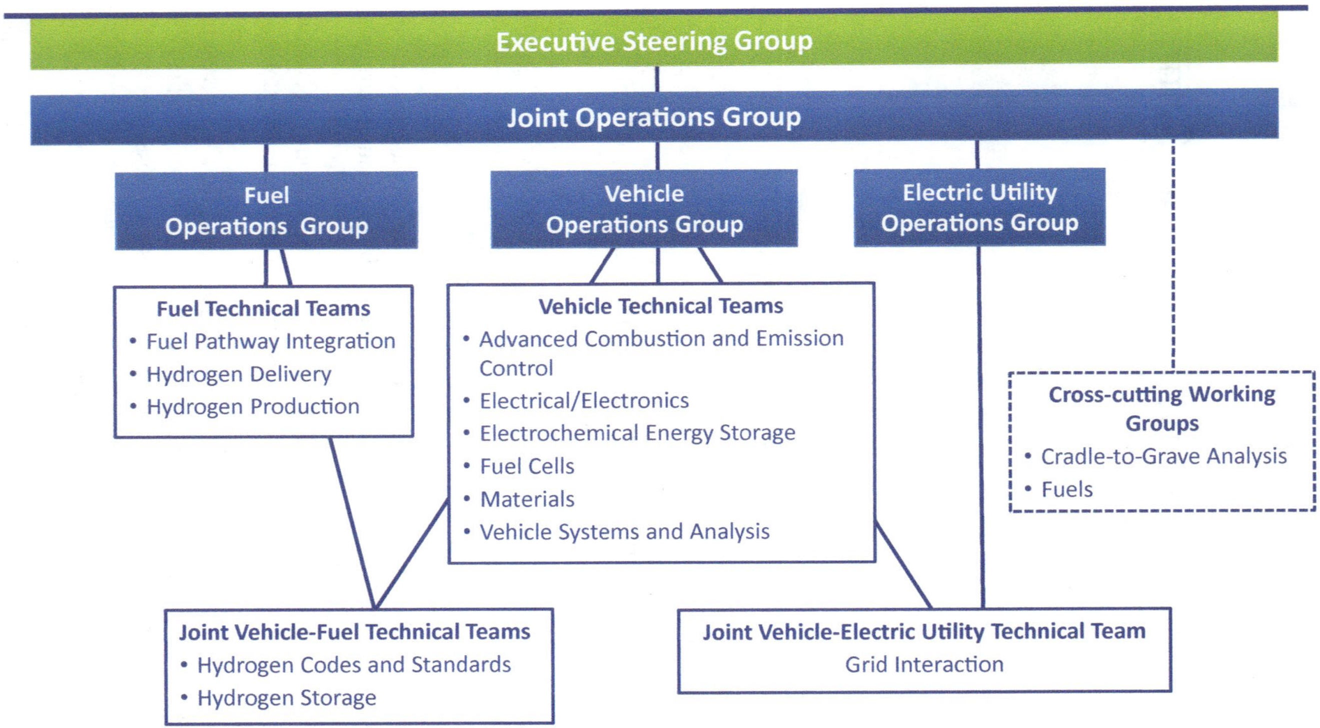

FIGURE 2-1 Organizational structure of the U.S. DRIVE Partnership.
NOTE: OEM, original equipment manufacturer. Note that a target setting task force discussed later in this chapter is not part of the formal organizational structure.
SOURCE: Cooper (2016b).

changes in ESG leadership) is the addition of the crosscutting working groups. The addition of the crosscutting cradle-to-grave (C2G) analysis working group in particular is directly responsive to a number of prior NRC recommendations. For example, as will be discussed later, Recommendation 2-1 in the NRC Phase 4 report called for a "portfolio-based strategy based on overall systems analysis performed by a proactive vehicle systems and analysis technical team and fuel pathway integration technical team" (NRC, 2013).

As with prior partnerships, the U.S. DRIVE Partnership also has industry-government technical teams responsible for setting technical and cost targets, as well as focusing appropriate research and development (R&D) on the candidate subsystems. Most of these technical teams focus on specific technical areas, but some, such as the hydrogen codes and standards technical team and the vehicle systems analysis technical team (VSATT) focus on crosscutting issues. A technical team consists of scientists and engineers with technology-specific expertise from the automotive companies, energy partner companies, utility industry companies, and national laboratories, as well as DOE technology development managers. Team members may come from other federal agencies if approved by the appropriate operations groups. A technical team is responsible for developing R&D plans and roadmaps, reviewing research results, and evaluating technical progress toward meeting established research goals. Its discussions are restricted to nonproprietary topics.

The U.S. DRIVE Partnership has expanded its outreach compared with the FreedomCAR and Fuel Partnership by including associate members from non-partner organizations. These associated memberships are for 3-year renewable terms. All but two technical teams (the electrochemical energy storage technical team and the materials technical team [MTT]) have an associate member. These associate members bring additional technical expertise and knowledge to the technical teams.

Furthermore, each U.S. DRIVE partner is involved in numerous other collaborations. U.S. DRIVE activities benefit greatly from these related outside efforts, as partners bring their knowledge and connections to the table as appropriate. Some examples of these other collaborations include Clean Cities, EERE Bioenergy Technologies Office (BETO), the 21st Century Truck Partnership (21CTP), and the Coordinating Research Council, as well as the Advanced Research Projects Agency-Energy and the Office of Basic Energy Sciences.

The various vehicle technical teams focus on advanced combustion and emission control, electrochemical energy storage, electrical/electronics, fuel cells, and materials, in addition to vehicle systems and analysis (see Figure 2-1). The three fuel technical teams address fuel pathway integration, hydrogen production, and hydrogen delivery. There are two joint vehicle-fuel technical teams connecting the fuel teams and the vehicle teams regarding hydrogen: the onboard hydrogen storage team and a hydrogen codes and standards team. Utility interface issues are handled by the grid interaction technical team. Finally, there are two crosscutting working groups, which are less formal than technical teams, have no roadmap or specific technical targets, but exist to offer structure to technical experts to convene and discuss particular topics of interest. The fuels working group is currently focused on low-carbon combustion fuels, and the new C2G working group is discussed in more detail later.

Within the DOE, primary responsibility for the U.S. DRIVE Partnership rests with the EERE. The two main program offices within EERE that manage the Partnership are the Vehicle Technologies Office (VTO) and the Fuel Cell Technologies Office (FCTO) (see Appendix B for an EERE organization chart). The focus of the VTO is on advanced technologies for clean, high-efficiency vehicles. Included in its portfolio are advanced combustion engine R&D, batteries and electric drive, vehicle systems, materials technology, fuels and lubricants, and outreach and analysis. In addition to R&D for light-duty vehicle technologies, the VTO also works with technologies applicable to medium- and heavy-duty vehicles through the 21CTP. There is considerable overlap between those elements of the VTO portfolio that relate to U.S. DRIVE and those that relate to 21CTP.

The mission of the FCTO is "to enable the widespread commercialization of hydrogen and fuel cell technologies, which will reduce petroleum use, greenhouse gas (GHG) emissions, and criteria air pollutants, and will contribute to a more diverse energy supply and more efficient use of energy" (Satyapal, 2016). The FCTO funds R&D activities on fuel cells, hydrogen fuel, manufacturing and distribution, and technology validation.

The U.S. DRIVE Partnership focuses on communication between the original equipment manufacturers and both the VTO and the FCTO at DOE, and the DOE laboratories with the principal objectives to do the following:

- Accelerate progress, discuss precompetitive issues among peers in the technical community, address technology-specific R&D needs, identify possible solutions, and evaluate progress toward jointly developed goals;
- Minimize duplication of efforts between government and industry;
- Ensure industry communicates its needs via the DOE R&D target setting process; and
- Remain focused on high-risk barriers to technology commercialization.

Some activities that are not part of U.S. DRIVE but that are related to the FCTO focus are not within the FCTO or even EERE. The Office of Fossil Energy has supported the development of technologies to produce hydrogen from coal and to capture and sequester carbon dioxide. The Office of Nuclear Energy has in previous years supported research into the potential use of high-temperature nuclear reactors to produce hydrogen, while BES supports fundamental work on new materials for storing hydrogen, catalysts, fundamental biological or molecular processes for hydrogen production, fuel cell membranes, and other related basic science areas (DOE, 2004a,b).

Within the EERE there is also BETO, which is not part of the U.S. DRIVE Partnership. However, biomass is of interest to the Partnership, as one possible source both of hydrogen as well as of biomass-based liquid transportation fuels (e.g., ethanol or gasoline or diesel derived from biomass) and as part of a strategy to diversify energy sources for the transportation sector; thus there is cooperation between the Partnership and the BETO. The committee believes, as discussed in this report and as mentioned in the Phase 4 report (NRC, 2013), that improving internal combustion engine (ICE) vehicles using biomass-based fuels is an important part of the portfolio of vehicle technologies that needs to be addressed.

And now, the increased emphasis on vehicle electrification suggests that understanding the interface between electric vehicle technology and the electric utility sector is of even greater importance. This heightened need is reflected in such recent DOE initiatives as the Grid Modernization Initiative and the EV-Everywhere Grand Challenge, which reside in EERE, within the Office of the Under Secretary of Energy.

VEHICLE AND FUEL PORTFOLIOS

A long-term goal of the Obama administration's Climate Action Plan and of the DOE's EERE was to "cut the Nation's greenhouse gas emissions by 17 percent below 2005 levels by 2020, 26-28 percent by 2025 and 83 percent by 2050" (Sarkar, 2016). This includes all sectors of the economy, but to achieve such a goal will require that light-duty vehicles achieve significant reductions in petroleum use and corresponding GHG emissions. Prior goals directly related to the technologies of interest to the Partnership included these: "Invest in developing electric vehicles technologies enabling one million electric drive vehicles on the road by 2015" and "reduce oil imports by 1/3 by 2025" (DOE, 2012), revised in February 2016 to "reduce net oil imports by half by 2020 from a 2008 baseline" (Sarkar, 2016). Another EERE goal, although not directly related to technologies under development by the U.S. DRIVE partners, is to "generate 80 percent of the Nation's electricity from a diverse set of clean energy sources by 2035" (Sarkar, 2016). If a large-scale penetration of battery electric vehicles (BEVs) or plug-in hybrid electric vehicles (PHEVs) takes place, then the goal of reducing GHGs significantly by 2050 will require an electricity production system that reduces such emissions significantly compared to the current U.S. electric power system.

As noted in the Phase 4 report, the main technology pathway options for reducing petroleum use and GHG emissions from light-duty vehicles are the following (NRC, 2013):

- *Reduce Vehicle Fuel Consumption:* Improve the fuel economy of light-duty vehicles through improved technologies, hybridization, lightweighting, and other vehicle design approaches in order to reduce both the amount of petroleum used per mile of travel and the associated GHG emissions.

- *Use Non-Petroleum-Based Liquid Fuels in Internal Combustion Engines (ICEs):* Use alternatives to petroleum-derived gasoline and diesel fuels in ICE-powered vehicles. Such fuels could include various alcohols (such as ethanol, methanol, or butanol) derived either from such nonpetroleum feedstocks as coal, natural gas, biomass, or garbage, or "synthetic" gasoline or diesel fuel derived from these feedstocks. The particular feedstocks and technologies used for the fuel production will determine the extent to which GHG emissions are reduced throughout the full fuel cycle.
- *Use Natural Gas in ICEs:* Use natural gas in ICE-powered vehicles. This reduces GHGs as compared to those from petroleum-based fuels, but the reduction in GHGs achieved will be less than for fuels that could be derived from non-carbon-based feedstocks or carbon-neutral biomass.
- *Use Hydrogen in ICEs or Fuel Cells:* Hydrogen can be used in either an ICE or a fuel cell. Much of the work by DOE in the Partnership has focused on developing better fuel cells and technologies for hydrogen production. If hydrogen is produced with low GHG emissions, the full fuel cycle can have a low GHG footprint.
- *Use Electricity in BEVs or PHEVs:* A BEV would use no other energy source on board the vehicle except for electricity from a battery, and a PHEV would travel some distance on electricity but would also have an ICE that would use fuel. Both types of vehicle would obtain the electricity from the electric power system, and their GHG emissions would depend on the extent to which the electric power grid is decarbonized, the number of miles that the vehicles could travel on electricity alone, the feedstock used for the production of the fuel used in the ICE on the PHEV, and the overall design of the vehicle for energy-efficient operation.

THE ROLE OF THE FEDERAL GOVERNMENT

As noted in the recent NRC (2015) report on a review of the 21CTP and also the prior NRC report on U.S. DRIVE (NRC, 2013), the role of the federal government in R&D varies depending on the administration and the Congress and the issues that they deem important for the nation to address. An extensive economics literature on the subject points to the importance of R&D to promote technical innovation, especially for research for which the private sector finds it difficult to capture the returns on its investment; this is especially true for basic research, the results of which can be broadly used. Such innovation, if successful, can foster economic growth and productivity, with improvements in the standard of living (Bernanke, 2011). Furthermore, in the energy area, the government generally has to confront issues of national security, environmental quality, or energy affordability. Many of these issues are addressed through policy initiatives or regulations, which place a burden on private firms to achieve. Thus, there is a role for the federal government in supporting R&D, not only to help the private sector achieve these policy goals but also to help U.S. firms remain competitive in the face of international competition.

The committee believes that the federal government plays an important role in the development of technologies that can help to address government policies and regulations aimed at reducing emissions and fuel consumption from light-duty vehicles. Such efforts as the U.S. DRIVE Partnership and the 21CTP are examples of public-private efforts to support R&D and to develop advanced technologies for vehicles. As noted by the NRC (2013), public-private partnerships generally include a variety of efforts (fundamental research, development, demonstration, and in some cases deployment). The federal government is well equipped to support fundamental and applied research and technology development through the national laboratories and universities, while industry can focus on product development and deployment. The importance of having government–industry collaboration is that the private sector can help to transform improvements from research into cost-effective and marketable products. Generally, the government contracting that is engaged in with the private sector is cost shared, and those research contracts more closely associated with fundamental or basic research will have a majority of federal funding, whereas contracts with a strong development or product component will have significant support from the private sector. Both U.S. DRIVE and 21CTP fall under the Energy Policy Act of 2005, which requires a minimum 20 percent cost share for R&D projects and a minimum 50 percent cost share for demonstration and commercial application projects. In its recommendations in each of the technical areas, the committee has considered what activities are precompetitive and are most appropriate for U.S. DRIVE and federal government support. Implicit in all of the recommendations that relate to the support of additional research, the committee believes that the federal government has a role in R&D.

TARGET SETTING PROCESS

Prior reviews of the U.S. DRIVE Partnership have been critical of both the target setting process and the decision-making process, due to the lack of an overall total vehicle systems analysis approach. The Partnership has taken steps to improve both these processes. In 2015 the Partnership established a target setting task force (TSTF), which operates in concert with the VSATT and the new C2G analytical working group. This task force established a four-step process for setting research targets. First, technical team inputs are used to define a virtual vehicle, using Autonomie software (developed by Argonne National Laboratory) for three different vehicle segments. Then a comparative cost metric is used to identify values of research target metrics that enable cost and performance parity at the vehicle level. Given this analytical context, teams can set or adjust targets as desired. This work is all performed by consensus.

The current research targets, all by 2020, are as follows:

- Electric vehicle full battery pack at a cost of $125 per kilowatt hour;
- Electric traction drive system at a cost of $8 per kilowatt;
- Automotive fuel cell system at a cost of $40 per kilowatt;
- Onboard hydrogen storage system at a cost of $10 per kilowatt hour;
- 20 percent improvement in engine efficiency, compared to 2010 baseline;
- 18 percent glider[2] mass reduction, relative to comparable 2012 vehicles, at a cost of $5 per pound saved; and
- Hydrogen fuel-related target that is DOE target of $2 to $4 per gallon gasoline equivalent.

During the committee's study in 2016, these targets were under review with the JOG and ultimately with the ESG and are expected to be updated. Furthermore, the teams are working to extend the target horizon to 2025, reflecting both future technology potential and in particular the recent volatility in oil prices.

The committee applauds the creation of the C2G working group and the TSTF, as well as the adoption of a more robust target setting process based on total system analysis.

Prior reviews have found that systems analysis has been applied very effectively at the subsystem or micro level, but that overall total systems analysis guiding high-level Partnership direction was lacking. These recent changes in approach appear to address these criticisms and are most welcome.

PARTNERSHIP DECISION MAKING

The Partnership has steadily evolved from PNGV, which had an explicit budget and portfolio, a robust go/no-go decision-making process with downselects, and specific hardware deliverables, through the FreedomCAR and Fuel Partnership to U.S. DRIVE, which has none of these.

Consequently, successive NRC review committees have struggled with the topic of decision making within the Partnership. This results from a number of factors, including the following:

- The ESG met rarely in the past, and even now meets only annually.
- The Partnership itself has no budget.
- It is not completely clear which projects actually reside within, or are associated with, the Partnership.
- The Partnership does not advise the federal government.

[2] The glider is the vehicle structure excluding the power train. Historically, weight reductions are easier to achieve in that part of the vehicle.

- The Partnership Plan states that each U.S. DRIVE partner makes its own decisions regarding its own funding of projects and programs according to its own internal policies.
- Similarly, each partner directs and manages its own projects and programs according to its own requirements.
- DOE has multiple inputs to its project portfolios, of which U.S. DRIVE is only one of many. See Figure 2-2.
- Each industry partner is at a different stage in development of relevant technologies.

In an effort to reconcile this ambiguity, the committee sought guidance from the Partnership, which provided the following clarifications.

The Partnership describes its activity as technical information exchange: the technical teams develop technology roadmaps that include Partnership research targets as well as a host of cascading targets and other requirements. These are all developed by team consensus. DOE relies on the U.S. DRIVE roadmaps and targets for guiding its research strategy and setting requirements for its R&D projects selected through open and competitive funding opportunities as well as its direct-funded national laboratory work. Roadmaps align with key DOE documents, including multiyear program plans and the EERE strategic plan.

In response to a committee request for a list of projects considered part of U.S. DRIVE, DOE provided a list of DOE projects that are associated with meeting U.S. DRIVE goals. However, this list contains many projects also listed as part of, for example, 21CTP, and are not exclusive to U.S. DRIVE. They appear to be, in fact, almost all of the projects handled by EERE that relate in some way to U.S. DRIVE and, as noted earlier, reflect input and guidance from U.S. DRIVE, but they are not directed or managed by the U.S. DRIVE. Conversely, in the Partnership response to the NRC Phase 4 Recommendation 3-17, and in the MTT presentation to this committee on June 22, 2016, some 13 and 12 (respectively) EERE carbon fiber projects were listed, whereas the project list submitted by DOE shows only two. The primary mechanism for assessing these EERE projects is the DOE Annual Merit Review (AMR) and not U.S. DRIVE, although of course many members of the U.S. DRIVE technical teams participate in the AMR.

The precise nature of decision making within the Partnership is apparently occasionally confusing to the partners themselves: for example, several presentations to the committee referenced go/no-go decisions, as does DOE's statement of task to the committee, but the DOE has stressed that the Partnership does not make go/no-go decisions.

Overall, as noted earlier and detailed in the technical sections of this report, the Partnership has an increasingly robust consensus process for developing goals and targets, and for providing guidance and input to DOE to help and inform the management of relevant DOE projects, and this process benefits greatly from the recent addition of overall strategic analysis. However, the supervision of those

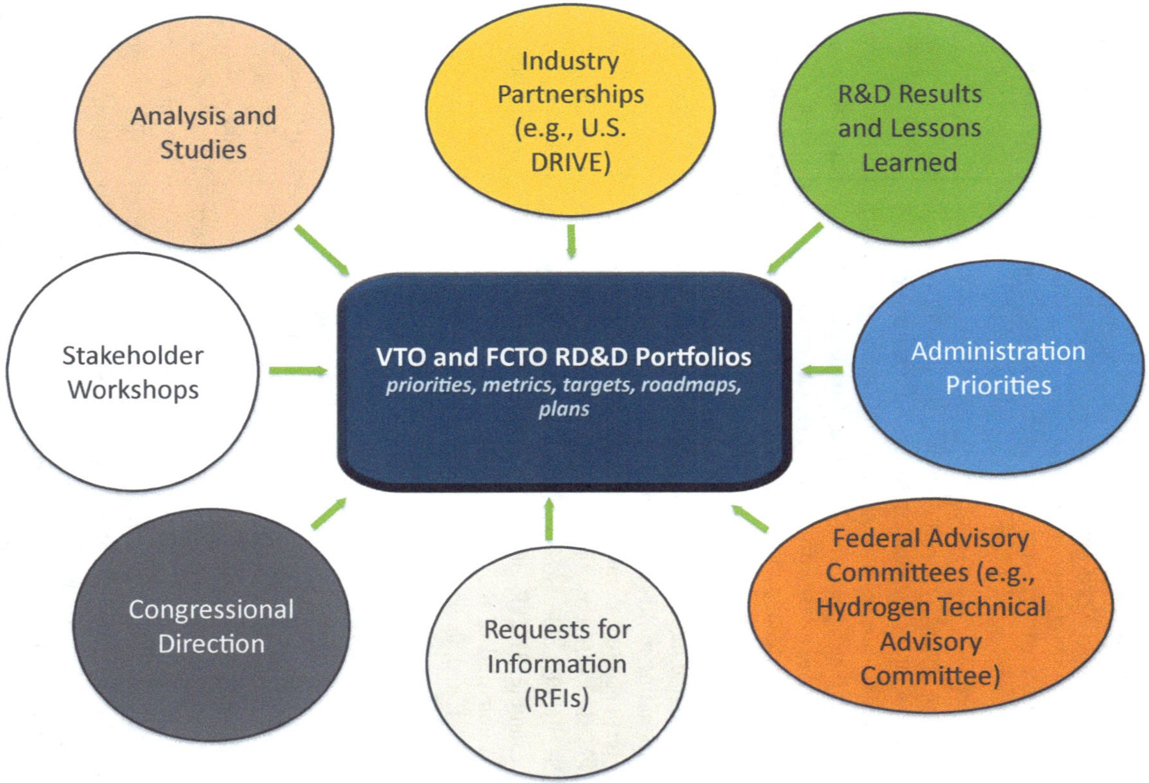

FIGURE 2-2 Inputs to DOE programs.
SOURCE: Cooper (2016a).

projects and the decisions made within them are a DOE EERE responsibility and not that of the Partnership.

The Partnership points to the type of decisions that *are* made by the Partnership as being, for example, focused within the portfolio, such as a decision (by the ESG) to emphasize work on low-carbon fuels.

Given these limitations on the scope of decision making within the Partnership, particularly limitations due to the Federal Advisory Committee Act, the committee feels that the processes applied within the technical teams are appropriate and that the Partnership has successfully fulfilled its mandate for technical information exchange.

RESPONSE TO NRC PHASE 4 RECOMMENDATIONS

The NRC Phase 4 report made four recommendations relating to program management and decision making: Recommendations 2-1 (also S-2), 2-2, 2-3, and 5-1 (also S-1). Those recommendations are reproduced here, together with the Partnership response and this committee's assessment of the response.

> ***NRC Phase 4 Recommendation S-2 and 2-1.*** The U.S. DRIVE Partnership should adopt an explicitly portfolio-based R&D strategy to help DOE to balance the investment among alternative pathways along with the more traditional reviews of the progress of individual pathways. Furthermore, this portfolio-based strategy should be based on overall systems analysis performed by a proactive vehicle systems and analysis technical team and fuel pathway integration technical team.

> ***Partnership Response.*** The Partnership supports a portfolio-based strategy and has developed an analysis-based mechanism for evaluating the potential benefits of technology pathways across its portfolio. Following a recommendation from the NRC's Phase 3 report and approval from the U.S. DRIVE Executive Steering Group, the Partnership established a cross-cutting cradle-to-grave analysis working group. The working group is tasked with examining the total energy use and greenhouse gas emissions of different pathways, including both fuel and vehicle manufacturing (including recycling) cycles to enable a comprehensive understanding of technology options within the U.S. DRIVE portfolio. The analysis includes pathways not currently included in the Partnership portfolio as well, such as various biofuel pathways and natural gas. The working group draws expertise from the fuel pathway integration and vehicle systems analysis technical teams, as well as additional analytical expertise from outside of those teams, and presented its work to the U.S. DRIVE Executive Steering Group in October 2012 and June 2013 for guidance on a path forward to help inform decision-making.

Committee Assessment of Response S-2 and 2-1. The committee feels that this response, and the corresponding actions taken by the Partnership, particularly the formation of the C2G working group, are fully responsive to the prior recommendation. The initial work by C2G on different vehicle-fuel pathways, published by Argonne National Laboratory and presented to the committee on June 22, 2016, is most impressive (this is discussed further in Chapter 3). It is important that this work and follow-on activity by the three analysis groups (C2G, VSATT, and the fuel pathway integration technical team) be used to shape the overall DOE

EERE portfolio. This includes transitioning the C2G working group to a "permanent integrated systems analysis technical team," a proposal under discussion by Partnership leadership in August 2016.

> ***NRC Phase 4 Recommendation 2-2.*** The Executive Steering Group (ESG) should meet regularly and provide the necessary guidance and leadership in developing strategy and programs to meet goals for the reduction of greenhouse gases and petroleum dependence. Furthermore, the ESG should insist that all analyses conducted by and for the U.S. DRIVE Partnership reflect the system-wide full lifecycle.
>
> ***Partnership Response.*** The Partnership agrees with this recommendation. The ESG maintains responsibility for high-level technical and management priorities as well as Partnership policy decisions. This duty can be accomplished only if ESG members are engaged and provide the necessary guidance and leadership for developing strategies and programs to meet Partnership goals. Although there was a period of time in which the ESG did not meet, since the formation of U.S. DRIVE in May 2011, the ESG has met three times and planning has begun for the next meeting in early 2014. It is also important to note that ESG members remain engaged in Partnership activities throughout the year through their staff on the Joint Operations Group (JOG) and technical teams, each of which meet monthly.

Committee Assessment of Response 2-2. While the increased engagement by the ESG is welcomed by the committee, annual meetings are barely adequate to provide the desired level of overall guidance and focus and meetings at least quarterly would seem more appropriate.

> ***NRC Phase 4 Recommendation 2-3.*** The U.S. DRIVE Partnership should continue its inclusion of innovative supply-chain companies and should expand this approach to emerging entrepreneurial companies with relevant technological capabilities. When new, entrepreneurial ventures are being considered for associate membership, the committee recommends a systematic vetting process much like the "due diligence" process of venture-capital investors.
>
> ***Partnership Response.*** The Partnership agrees with the recommendation, as it aligns with the intent of its "associate membership" concept. The selection process for associate members is described in the U.S. DRIVE Tech Team Guidebook as well as U.S. DRIVE Partnership Plan. Associate members can be any entity that a U.S. DRIVE technical team believes will bring sufficient expertise, capability, and contribution to the team's efforts. Decisions are made at the team level, by consensus among the participating U.S. DRIVE partners on the team, and with consideration of the following factors:
>
> - Availability to participate in meetings and contribute to the team's work for a three-year term (or length of term as determined by the team).
> - Technical capability and personal experience in the field, including (but not limited to) years of experience, management of relevant programs, publications and patents, record of bringing innovations to market, etc.
> - Ability to work well in a group environment—including the ability to contribute to group discussion as well as the ability to engage in healthy dialogue and still get along well with others.
>
> The Partnership also recognizes the importance of evaluating potential conflict of interest issues, and associate membership parameters such as term-limited participation allow teams to engage a number of organizations over time and as dictated by their needs. Each partner may also have its own internal vetting process that informs the team's consensus decision.

Committee Assessment of Response 2-3. The committee appreciates the Associate Member outreach and the rules under which the U.S. DRIVE teams and working groups engage supply chain partners. The committee thinks, however, that this approach will present a cultural and economic challenge to early stage companies: those funded under the DOE Small Business Innovation Research (SBIR) program, for example. More university researchers with appropriate expertise might be considered as Associate Members of the technical teams.

These challenges arise from a mismatch in objectives. The U.S. DRIVE Partnership Plan of April 2016 states that the objective of Associate Membership is to provide the current partners with "additional experts with diverse perspectives, including technical knowledge uniquely relevant to a specific technical area." In contrast, early stage companies chiefly seek market opportunities and capital investment. To the extent that these are not offered within the Partnership, its attractiveness to early stage companies will diminish.

This means that Partnership meetings must include a reasonably foreseeable opportunity for entrepreneurs to advance their ventures (no guarantees needed). For example, the U.S. DRIVE partners might consider hosting an invitation-only venture forum for selected SBIR companies. The committee notes that the U.S. Council for Automotive Research has some successful experience with these.

> ***NRC Phase 4 Recommendation S-1 and 5-1.*** The Executive Steering Group should be engaged to set targets for the U.S. DRIVE Partnership that are consistent with the objectives of reduced petroleum consumption and greenhouse gas emissions, and U.S. DRIVE should conduct an overall review of the Partnership portfolio, both for the adequacy of the R&D effort to achieve the targets and for focus on the mission of supporting longer-term, higher-risk precompetitive activities in all three potential primary pathways.
>
> ***Partnership Response.*** As part of its responsibility for high-level technical and management priorities, the Executive Steering Group maintains approval authority over Partnership targets for the U.S. DRIVE technology portfolio. Following an intensive, analysis-based process, the ESG approved new U.S. DRIVE Partnership targets that were included with the U.S. DRIVE Partnership Plan in March 2013. These targets align with U.S. DRIVE goals for the technology areas in its portfolio, each of which contributes to petroleum and greenhouse gas (GHG) emission reductions in the transportation sector. Technical team roadmaps include multiple cascading and other targets, as well as requirements for each technology area, which align with the high-level Partnership targets and goals in the Partnership Plan.
>
> Each technical team conducts a portfolio review, and individual project reviews at the technical team level provide the opportunity to examine and discuss progress toward (and challenges to) achieving Partnership and technical targets. Teams regularly report results to the Partnership's Joint Operations Group (JOG) members. In addition, the Partnership's cradle-to-grave (C2G) activity is studying the petroleum and greenhouse gas reduction potential of pathways. It regularly reports the status and results of its activity to the JOG and has updated the ESG at its meetings in October 2012 and June 2013. The group plans to publish this work on the DOE web site.

Committee Assessment of Response S-1 and 5-1. These actions, particularly the creation and deployment of the C2G working group and the new target setting approach, are fully responsive to the recommendations, with the earlier caveats regarding timely engagement by the ESG and the transition of C2G to a permanent technical team.

FINDINGS AND RECOMMENDATIONS

Finding 2-1. The committee finds that the response to prior recommendations regarding management of the Partnership, particularly the creation of the target setting task force and the cradle-to-grave (C2G) working group, and the adoption of a portfolio-based strategy are welcome improvements, and the Partnership is well managed. However, the increased engagement by the Executive Steering Group has improved from unacceptable to barely adequate. Furthermore, the Partnership currently regards the C2G working group as only a "temporary, task-specific" group.

Recommendation 2-1. Now that there is a more robust target setting task force and cradle-to-grave working group, the Executive Steering Group should meet more regularly than annually (perhaps at least quarterly) and participate directly in the portfolio analysis and target setting process for revised 2020 and new 2025 goals. Furthermore, the recently published cradle-to-grave study on vehicle-fuel pathways and follow-on work by the target setting task force and cradle-to-grave working group should be used proactively and specifically to help shape the overall Office of Energy Efficiency and Renewable Energy portfolio, and the cradle-to-grave working group should be transitioned from temporary to permanent status.

Finding 2-2. Given the reality that the Partnership does not direct or manage DOE-funded programs, overlaps with other DOE programs, and has no budget, there remains considerable ambiguity over the precise scope of the Partnership and its relationship with other DOE activities.

Recommendation 2-2. The Partnership is urged to provide more transparency and clarity regarding those Department of Energy projects deemed wholly or partly within the U.S. DRIVE portfolio and the achievements truly attributable to the Partnership.

REFERENCES

Bernanke, B.S. 2011. Promoting research and development: The government's role. *Issues in Science and Technology* 27(4). http://www.Issues.org/27.4/bernanke.html.

Cooper, C. 2016a. Inputs to DOE Programs. E-mail communication to the committee (pers. comm.), March 4.

Cooper, C. 2016b. "U.S. DRIVE Overview Presentation." Presentation to the Committee on the Review of the Research Program of the U.S. DRIVE Partnership, Phase 5, Washington, D.C., February 3.

DOE (U.S. Department of Energy). 2004a. *Basic Research Needs for the Hydrogen Economy: Report of the Basic Energy Sciences Workshop on Hydrogen Production, Storage, and Use, May 13-15, 2003*. Washington, D.C.: U.S. Department of Energy, Office of Science. http://science.energy.gov/~/media/bes/pdf/reports/files/Basic_Research_Needs_for_the_Hydrogen_Economy_rpt.pdf.

DOE. 2004b. *Hydrogen Posture Plan: An Integrated Research, Development and Demonstration Plan*. Washington, D.C.: U.S. Department of Energy. http://www.4uth.gov.ua/usa/english/tech/energy/hplan.pdf.

DOE. 2012. *Department of Energy FY2013 Congressional Budget Request*. Vol. 3. February. DOE/CF-0073. Washington, D.C.: U.S. Department of Energy, Office of the Chief Financial Officer.

NRC (National Research Council). 2013. *Review of the Research Program of the U.S. DRIVE Partnership, Fourth Report*. Washington, D.C.: The National Academies Press.

NRC. 2015. *Review of the 21st Century Truck Partnership, Third Report*. Washington, D.C.: The National Academies Press.

Sarkar, R. 2016. "EERE Sustainable Transportation Overview." Presentation to the Committee on the Review of the Research Program of the U.S. DRIVE Partnership, Phase 5, Washington, D.C., February 3.

Satyapal, S. 2016. "Overview of the Fuel Cell Technologies Office." Presentation to the Committee on the Review of the Research Program of the U.S. DRIVE Partnership, Phase 5, Washington, D.C., February 3.

3

Light-Duty Vehicle Technologies and Fuels

INTRODUCTION

This chapter covers the technology areas funded by the U. S. Department of Energy (DOE) Vehicle Technologies Office (VTO) and Fuel Cell Technologies Office (FCTO), which support U.S. DRIVE (*D*riving *R*esearch and *I*nnovation for *V*ehicle Efficiency and *E*nergy Sustainability) goals and targets. These topics are advanced combustion engines, fuels and emissions, fuel cells and hydrogen (onboard storage, production delivery and distribution, safety codes and standards), electric propulsion (electric drive systems, power electronics, electrochemical energy storage, electricity as an energy source, and the grid interactions), structural materials, and cradle-to-grave (C2G) analysis and implications.

The task of the committee is not to provide a detailed technical review of each of the projects funded by DOE, which are those that contribute to U.S. DRIVE Partnership goals, as this is accomplished through the DOE Annual Merit Review (AMR) meeting, but rather to review progress and program directions and an assessment of ongoing research activities, as well as how effective the Partnership is in identifying precompetitive research and development (R&D) activities that lead to the achievement of Partnership goals.

To this end, each of the subsections addressed in this chapter provides a brief background of the technology and its importance to the goals of the Partnership; a current status vis-à-vis Partnership goals, targets, and timetables; an assessment of progress and key achievements and program directions; a DOE budget overview; significant barriers and issues that need to be addressed; the response to recommendations from the National Resource Council (NRC) Phase 4 review (NRC, 2013); the appropriate federal role; and findings and recommendations (see Chapter 1 for the full statement of task).

Internal combustion engines (ICEs) have been the dominant vehicle power plant since the inception of mass-produced personal transportation and can in many respects be considered a mature technology. Yet there is still potential for their improvement in terms of lower tailpipe emissions, reduced fuel consumption, and a lower greenhouse gas (GHG) total footprint. Given the number of vehicles in this class, activities related to advanced combustion technologies of vehicles using liquid fuels and powered by internal combustion engines will be a major part of DOE's technology pathway to achieving an 83 percent reduction in GHG emissions by 2050 (Sarkar, 2016). Efforts to reduce the fuel consumption of ICE power trains are directed at improving the engines' peak efficiency and being able to achieve that maximum efficiency over the engines' entire operational range.

In order to fully meet the Partnership goals and current and future standards, additional approaches will be required. Electric drive vehicles, while still a small part of the market, have emerged as strong candidates for meeting these requirements, provided that a green electric grid evolves in parallel. The introduction of hybrid electric vehicles (HEVs) in 1999 initiated a significant move toward vehicle electrification, with many HEV options now commercially available. Advances in battery technology have enabled the introduction of plug-in hybrid electric vehicles (PHEVs) and battery electric vehicles (BEVs), which reduce GHG emissions even further. As new battery materials evolve, greater range, reduced charging times, and reduced cost may be achieved. Hydrogen fuel cell vehicles (HFCVs) provide an attractive electric alternative that addresses both refueling time and range but are currently limited by cost and the lack of a hydrogen refueling infrastructure.

It is worth noting that during this review Toyota, Hyundai, and Honda have made available within the United States a limited number of fuel cell vehicle sales or leases to the general public. General Motors, a U.S. DRIVE Partnership member, has reported they plan on a 2020 rollout of their latest fuel cell vehicle, following years of extensive and successful road testing of their first-generation HFCV, the Equinox. The development and deployment of roadworthy HFCVs is a major accomplishment and one that will help to identify remaining technical, cost, manufacturing, and infrastructure challenges. Though the cars are still in the late stages of development, the fact that these cars have advanced to this point is due in part to R&D coordination by the Partnership and its prior organizations, as well as from decades of funding of pertinent research projects by the DOE and Partnership members.

The lack of a hydrogen infrastructure remains a significant challenge that will impact the degree of acceptance and success of the fuel cell electric vehicle, especially in the early years. This not only includes production and delivery of hydrogen but also cost on a gallon of gas equivalent (GGE) basis, if such vehicles are to be competitive with other electric vehicles as well as current internal combustion technologies. Furthermore, the generation of hydrogen from renewable sources must continue to be researched, developed, and ultimately adopted if the

HFCV GHG emissions are to be reduced below other vehicle technologies. The near-zero emissions potential when an HFCV is fueled with hydrogen generated from renewable energy sources would be a significant breakthrough in the quest to limit GHG emissions from the transportation sector.

Finally, the deployment of lightweighting materials and components and more efficient electric power trains will contribute to increased vehicle range and decreased emissions, both being key goals of the Partnership.

ENGINES AND EMISSION CONTROLS

Introduction

Regulations and future targets for GHG emissions have instilled new urgency to reducing the fuel consumption and total GHG footprints of the nation's vehicle systems. This is spurring further development of ICEs, their associated power trains, aftertreatment systems, and the characteristics of the fuels used to carry the energy into the engine. The advanced combustion and emissions control technical team (ACECTT) guides the effort of improving the performance of internal combustion engines, fuels, and aftertreatment systems for the U.S. DRIVE Partnership.

In assessing the relevance and accomplishments of the engine, fuel, and aftertreatment research programs within the ACECTT portfolio, it is important to understand that the improvement path for efficiency in ICEs is not without limits. There are fundamental and practical limits to the efficiencies that can be obtained, just as there are for fuel cell and battery electric systems. In March 2010 DOE hosted a colloquium on Transportation Combustion Engine Efficiency at USCAR to discuss these limits (Daw et al., 2010). The purpose of the colloquium was to engage academics, researchers, and developers in a review of the underlying thermodynamics applicable to ICEs, to identify the unavoidable losses (irreversibilities, i.e., exergy destruction) that must be accepted when producing power from chemically reacting systems, and to opine as to what might be maximum practical efficiencies achievable with internal combustion engines.

Although a definitive number was not agreed upon at the colloquium, the outcome of the meeting indeed did provide stretch goals presented to the technical community—for example, the targets set for the Super Truck programs. For heavy-duty diesel engines, brake thermal efficiencies on the order of 55 percent are in the vicinity of maximum practical efficiencies for a stand-alone ICE. Peak engine efficiencies are typically achieved at high-load operating conditions. For smaller engines, because of less favorable geometric characteristics such as surface area to volume ratios and friction factors, the peak efficiency is less than for larger engines; however, as with the larger engines the highest efficiencies are achieved at high load. In general the efficiency of the engine decreases as its operating condition moves away from high loads into lower-load operating regimes.

Despite the fact that some engines are approaching their practical maximum efficiencies, there is still room for improvement. Big large-bore, long-stroke engines have peak efficiencies that indeed are very near their practical limits. There is little opportunity to improve their peak efficiencies further. Engines used in light-duty vehicles will have lower practical maximum efficiencies than big engines, and the current values of their peak efficiencies are not as close to their practical maximums. The opportunities for improvements in smaller engines lie in activities to push their peak efficiencies closer to their practical maximum limits and efforts to achieve these higher efficiencies over a large portion of the engine operating regime.

The ACECTT has set program targets of efficiency gains of 20 percent by 2020 relative to a 2010 baseline engine when operating over a prescribed set of operating conditions. Since the efficiency typically drops as the operating condition moves away from the point of maximum efficiency—that is, moves to lower loads—there is much that can be done to improve the overall cycle efficiency. If the high-load efficiency of the engine can be replicated at lower loads, the overall cycle efficiency of the vehicle will improve, which would result in significant reductions in fuel consumption and emissions. Furthermore, as engine evolution takes place, changes in combustion processes and technologies for overcoming practical constraints in these smaller engines—like being able to operate knock free at higher compression ratios—will improve the peak engine efficiency. These improvements in peak efficiency could then be promulgated over the entire operating cycle of the engine and act as an additional multiplier to improvements to the overall cycle efficiency. Incorporating technologies that accomplish these goals (ca. 2020) would occur via model year changes of vehicles and would thus penetrate the market at the rate of the vehicle fleet turnover. Given the magnitude of the vehicle fleet powered by ICEs, fuel consumption and emission reductions introduced through ICE improvements would be multiplied by millions of vehicles per year, so the reduction in fuel consumption and emissions would be significant.

As a transportation power plant the ICE is ideal for applications that demand sustained high power operation and long times or distances between refueling. However, a major challenge in using ICEs for mobility applications is that typical mobility duty cycles contain significant intermediate- and light-load operation, where the engine's efficiency drops to levels that can be considerably below its peak value.

Consequently, efforts to reduce the fuel consumption of ICE power trains are directed at improving the engines' peak efficiency, being able to achieve that maximum efficiency over the engines' entire operational range, and reducing the auxiliary, accessory, and friction loads that detract from the power delivered to the wheels. However, even the most efficient engine will still need exhaust gas aftertreatment to ensure that the criteria tailpipe pollutants are below regulated limits over the entire vehicle operating domain. So, the engine and the aftertreatment components must be developed as a system to achieve optimal performance. Accomplishing this requires understanding all the energy flows, transformations, and work-potential dissipations (exergy destruction) that occur in the power train—that is, starting from the fuel leaving the fuel tank to the power being delivered to the wheels and then guaranteeing that the aftertreatment system can function effectively at the exhaust conditions leaving the engine cylinder. Achieving improvements in these areas pushes the boundaries of the current understanding of almost all the physical and engineering sciences.

Pursuing such improvements is the focus of the engine combustion and aftertreatment research community. It is manifest in such activities as hybridization of the power plants, downsized boosted engines, variable valve actuation, cylinder deactivation, and advanced combustion processes. In each of these applications the technical community is working both to increase the peak efficiency and to bring high-load engine efficiencies to vehicle operation that does not require high load engine operation.

As the engine efficiency is increased more of the fuel energy is converted to work and there is less energy leaving the engine in the exhaust stream. The lower energy content of the exhaust thermodynamically translates into lower exhaust gas temperatures. This presents an additional challenge for the exhaust gas aftertreatment system. Not only is there need for higher effectiveness of the aftertreatment systems to reduce criteria pollutants, but also the aftertreatment systems need to achieve this effectiveness at lower temperatures.

In addition, the fuel cycle needs to be part of the focus. Biomass-derived fuels, either as blends or drop-in components, have the potential to reduce total life cycle (also referred to as C2G) GHG emissions. Research is under way to explore potential synergies between fuel refining processes and fuel characteristics selected to enhance the combustion process while trying to achieve optimal C2G GHG emissions for the system as a whole. For example, higher octane number fuels would allow engines to have higher compression ratios, which improves their efficiency and could eliminate the need for spark retard (a control strategy to avoid engine knocking that is detrimental to efficiency). However, the extent to which this can be done, and the implications in the trade-offs in GHG emissions between the additional processing and potential costs necessary to achieve the higher octane number during the fuel production versus the lower fuel consumption achieved during engine operation, needs to be determined.

Fuels will also play an important role in achieving the 2050 carbon dioxide (CO_2) reduction targets for the light-duty vehicle fleet by reducing the carbon footprint of the energy carrier itself. The advanced engine and combustion strategies under investigation by the ACECTT will enable significant CO_2 emission reduction; however, without an accompanying effort to reduce the carbon in the fuel it is unlikely that the 2050 CO_2 target can be met (Farrell, 2016). DOE started an initiative on co-optimization of fuels and engines or "Co-Optima" (DOE, 2016c) in fiscal year (FY) 2016 to address these challenges. Although the co-optimization initiative is a separate DOE effort from the U.S. DRIVE Partnership, it is providing funding support for DOE research projects that are considered as supporting the goals of U.S. DRIVE.

ACECTT Research

The U.S. DRIVE ACECTT mission statement reads as follows:

> Reduce petroleum dependence by removing critical technical barriers to the mass commercialization of high-efficiency, emissions-compliant internal combustion engine (ICE) powertrains.

Their activities focus on continued improvement of spark-ignition (SI) and compression-ignition engines working in conjunction with more effective aftertreatment systems, while continuing the development of more advanced, kinetically controlled, low-temperature combustion engine concepts.

The ACECTT activities are fundamentally dissecting and analyzing the energy flows within the engine-aftertreatment-power train system to address every energy transformation that occurs within the vehicle. This includes the energy flow in the fuel leaving the fuel tank, through its introduction into the cylinder, during the combustion and work extraction processes, through the aftertreatment system, and out the tailpipe. Attention is also directed at all energy flows that use work generated by the engine for functions other than driving the wheels—such as accessories, pumping and friction; or energy flows that leave the engine in a form other than work, such as heat transfer and exhaust flow. Maximizing efficiency and minimizing emissions over the operating map of the vehicle is a complex challenge that requires detailed understanding of the fundamentals that govern these energy transformations, the degradations of the work-potential of the energy associated with these transformations—that is, exergy destruction—as well as being able to measure them via sensors and developing advanced control systems to optimize the energy management for the entire system. As an enhanced understanding of these energy flows, and their associated exergy destruction, is developed, it enables researchers to determine which irreversibilities can practically be pursued for reduction, and the extent to which efficiency improvements are possible by perfection of the energy transformation in question. For example, the following have all been important in directing research and technology development activities: understanding the magnitude of the work that is lost due to the chemical reactions that release the chemical energy in the fuel, and realizing that this loss cannot be prevented with current technologies; or understanding the relative importance of the heat transfer from the engine during different portions of the combustion process; or being able to calculate the work that might be obtained by exhaust energy recovery systems; or understanding the trade-offs of different exergy destruction processes that lead to improved engine efficiency by keeping combustion temperatures low. These activities are made especially challenging because to gain the maximum benefit they need to be understood accurately for transient operation.

To address the interactions between fuels and the engine system U.S. DRIVE has established a Fuels Working Group (FWG). A working group is less formal than a technical team. It does not have a roadmap or defined technical targets but instead offers structure for technical experts to discuss topics of interest. The ACECTT, interfacing with the FWG, has sharpened the focus of fuels testing within the ACECTT engine testing programs. The ACECTT, the FWG, and USCAR have developed a well-to-wheels study fuel set that identifies fuel formulations with a range of research octane numbers (RONs), octane sensitivities, and biofuel energy content that might be representative of fuels that could be made available in the near term (Farenback-Brateman et al., 2016). These fuels now serve as a common base for the experimental spark-ignition ICE programs within the ACECTT portfolio. The activities promoted within the ACECTT by the FWG are closely aligned with the objectives of the newly formed DOE Co-Optima initiative.

The ACECTT understands that advanced spark- and compression-ignition engines will be dominant in the near-term mobility fleet. The potential of advanced low-temperature combustion systems, also referred to by the ACECTT as chemical kinetics-dominated combustion, has been demonstrated in the laboratory (Ra et al., 2012); however, in the February 2016 committee meeting the ACECTT indicated that chemical kinetics–dominated combustion was not considered a near-term technology. The technical team expects that the engine types used during the time period of focus for this program will be a mix of naturally aspirated hybrids and advanced downsized boosted architectures. That is, the ACECTT expects that in the near term the engines will have conventional four-stroke architectures with improved but relatively conventional spark-ignited flame propagation or diesel–type autoignition combustion systems. The ACECTT specifically did not suggest dates for near- and long-term time frames. Furthermore, the ACECTT believes that within each of these engine pathways, the combustion strategies will be one of the following:

- Spark ignition, where combustion is flame propagation dominated and dependent on premixing of the air and fuel prior to flame initiation;
- Compression ignition, where combustion processes are mixing and diffusion dominated; or
- Chemical kinetics dominated, low-temperature combustion (LTC).

Each of these combustion strategies depends on the characteristics of the fuels being used. Fuel characteristics such as resistance or propensity to autoignite and how that propensity changes with the conditions in the cylinder, burning velocity, tendency to form particulate matter, and heat of vaporization will impact the effectiveness of the combustion strategy in converting the energy in the fuel to power out of the engine. Through the Fuels Working Group, the ACECTT is also attempting to identify optimal fuel characteristics for the different combustion strategies and determining if the introduction of such fuels could be easily integrated with the current fuel infrastructure and legacy fleets. These research issues are also part of the DOE Co-Optima initiative; these are areas where Co-Optima funding is being directed toward projects that support U.S. DRIVE goals.

The ACECTT has identified specific strategies and research targets to promote progress toward their goals. For premixed flame dominated (SI) engines, pursuing dilute combustion is recommended. Combustion in near-term dilute SI engines will be dominated by the propagation of a flame front through reactants that are largely premixed. The dilution of the charge, either with excess air (lean) or with exhaust gas recirculation (EGR) promotes higher efficiency because the combustion temperatures are lower than those occurring in undiluted, stoichiometric[1] spark-ignited combustion. The lower combustion temperatures result in a ratio of specific heats, gamma, that is higher than would occur if the gas temperatures were higher. Thermodynamically this leads to a higher work extraction per unit of piston motion during the expansion stroke, as well as lower heat transfer and exhaust enthalpy losses. It is also beneficial because the cylinder-out oxides of nitrogen (NO_x) emissions are lower owing to the lower temperatures, so less NO_x reduction is needed from the aftertreatment systems. Particulate emissions are also low because the reactants are largely premixed, so particulate filtration requirements are less.

For the near-term engines, which are mixing and diffusion dominated (CI), the technical team advocates "clean diesel." In diesel engines the combustion is mixing controlled, where burning takes place in conjunction with the mixing of fuel and oxidizer. Low emissions from diesel combustion can be achieved by using EGR with advanced mixing and injection strategies, which must then be coupled with effective aftertreatment systems. Enhanced understanding of the interactions among the EGR, advanced injection strategies, and in-cylinder fluid motion is needed to promote combustion conditions that minimize the in-cylinder regions of high-soot and high-NO_x formation.

To move beyond the near-term engine combustion technologies ACECTT encourages continued fundamental development of chemical kinetics-controlled combustion, or LTC. LTC is used here as the name for the generic combustion process that is largely flameless, volumetric autoignition that is controlled by chemical kinetics. Control of LTC is achieved by staging the autoignition of different spatial regions of dilute (lean and/or EGR) fuel-air mixtures at temperatures below those necessary for rapid flame propagation: temperatures typically below approximately 1750°C but above approximately 1250°C. Many approaches are being investigated to achieve and control this staging of the autoignition, and this

[1] Stoichiometric combustion is when the mixture of fuel and air is chemically correct, such that there is just enough oxygen so that all the carbon and hydrogen in the fuel could be converted to carbon dioxide (CO_2) and water (H_2O), respectively, with no oxygen left over.

has led to an alphabet soup of acronyms: HCCI (homogeneous charge compression ignition), PPCI (partially premixed compression ignition), RCCI (reactivity controlled compression ignition), GCI (gasoline compression ignition), and so on. However, in all of these combustion modes the underlying objective is the same: one wants to establish a somewhat premixed mixture of fuel-vapor, air, and diluent that is encouraged to progress through the air-fuel mixture's autoignition chemistry in such a way that the mixture ultimately chemically reacts through a rapid sequence of autoignition events that occur volumetrically throughout the cylinder. Successfully creating this situation inside the combustion chamber depends on the state of the gases at the start of compression and the physical and chemical characteristics of the fuel. The litany of acronyms that has appeared in the literature is an indication of the different approaches that can be taken to achieving this combustion mode. The benefit of LTC is higher efficiency, for the same reasons cited earlier in the discussion of dilute gasoline engines, and very low NO_x and soot.

As the engine becomes more efficient, more of the fuel energy is being converted to shaft work, so less energy is leaving the engine as exhaust enthalpy. Consequently, the engine exhaust is at a lower temperature than it is from a less efficient engine. This presents an additional challenge for the aftertreatment system. Today's aftertreatment systems require catalyst bed temperatures in the range of 400°C to 600°C (Johnson, 2016). Industry is aggressively pursuing and making progress in lowering the light-off temperatures and temperature window of full functionality for the aftertreatment systems. Ideally one would like aftertreatment systems that are fully functional at cold start temperatures. For this reason the ACECTT is advocating continued research into perfecting aftertreatment systems that work at lower temperatures. This is needed for all aftertreatment systems: three-way catalysts, lean NO_x traps, selective catalytic reduction systems for NO_x, and particulate filters.

To benchmark progress and motivate research efforts the ACECTT has established research targets for 2020 for both engine efficiency improvements and exhaust emission levels. The high-level statement of the Partnership's research target, which was presented to the committee, is given as follows.

2020 Partnership Research Target: *A 20% improvement in engine efficiency, compared to a 2010 baseline. Engine concepts shall be commercially viable and meet 2020 emissions standards.*

The research targets have been broken down for the specific engine types found in the market today, projecting that near-term engines will be further developments of current state-of-the-art engines. These engine-specific research targets are shown in Table 3-1. The targets are stated in terms of improvements relative to a 2010 baseline for each engine type. As shown in Table 3-1, the technical team has highlighted the operating points for each of the different engine pathways

TABLE 3-1 Advanced Combustion and Emission Control Technical Team (ACECTT) Stretch Goals for 2020 for Different Engine Types Currently Being Used in the Light-Duty Mobility Market

		2010 Baselines				2020 Stretch Goals[a]		
Technology pathway	Fuel	Peak efficiency[b]	Efficiency[b] at 2-bar BMEP and 2,000 rpm	Efficiency[b] at 2,000 rpm and 20% of the peak load	2,000 rpm peak load[c]	Peak efficiency	Efficiency at 2-bar BMEP and 2,000 rpm	Efficiency at 2,000 rpm and 20% of the peak load
Hybrid application	Gasoline	38	25	24	9.3	46	30	29
Naturally aspirated	Gasoline	36	24	24	10.9	43	29	29
Downsized boosted	Gasoline[d]	36	22	29	19	43	26	35
	Diesel	42	26	34	22	50	31	41

[a] Entries in percent brake thermal efficiency (BTE) that are equal to 1.2 times the corresponding baseline BTE.
[b] Entries in percent BTE.
[c] Entries in bar of brake mean effective pressure (BMEP).
[d] Downsized boosted baseline engine used premium grade fuel and direct injection.

NOTE: Baseline is 2010 light-duty vehicle U.S. penetration (multivalve, 86%; port injected, 91%; variable valve timing, 86%; stoichiometric with three-way catalytic converter, 99% [for less than 8,500 lb gross vehicle weight]). Highlighted cell represents most relevant operating point for that technology pathway.
SOURCE: Solomon and Howden (2016).

that it feels are the most relevant. The highlights represent the most important operating conditions at which research engines of that type should be thoroughly evaluated. The research targets for emissions were established based on current and pending regulations, as shown in Figure 3-1.

At the request of the committee the U.S. DRIVE Partnership provided a list of the projects within DOE's advanced combustion engine research portfolio in which they are engaged. This list consists of 40 projects and is given in Table 3-2. The total DOE funding for FY 2015 for the projects listed in Table 3-2 was $29.02 million. The DOE funding allocation for these projects in FY 2016 decreased slightly to $27.5 million: several of the projects were completed in 2015 or early 2016, which is the primary reason for the respective differences in the funding levels.

The ACECTT has identified the fundamental areas (barriers) for which enhanced understanding would facilitate progress toward their research targets for each of the projects listed in Table 3-2. Rather than describe the specific barriers being addressed in each of the projects listed in the table individually a more general discussion is presented below, where the needed fundamental understandings are described in terms of the phenomena that currently limit a particular engine's performance. This discussion is presented in the context of the three individual combustion strategies, the development of enhanced computational fluid dynamic (CFD) capabilities, and the challenges of future aftertreatment systems. The entries in Table 3-2 have been color coded to show the connection of the projects in the table to the general descriptions of the challenges for the different combustion strategies, CFD development, and aftertreatment systems. If a project advances more than one category of the general description it is shown with multiple color codes. Further detail on the individual projects in the table can be found in the presentations for the VTO given at the DOE 2015 and 2016 AMRs with the project numbers contained in Table 3-2.

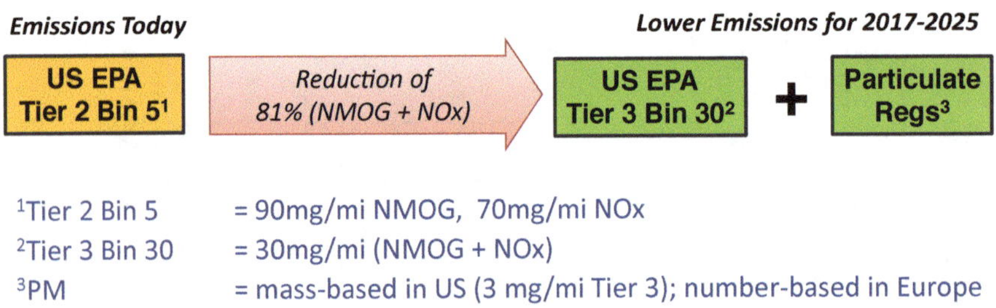

FIGURE 3-1 ACECTT (advanced combustion and emission control technical team) emission research targets. NOTE: NMOG, non-methane organic gas. SOURCE: Solomon and Howden (2016).

TABLE 3-2 List of Department of Energy Projects Addressing U.S. DRIVE Goals

Project ID	Presentation Title	Organization
ace010	Fuel Injection and Spray Research Using X-Ray Diagnostics	ANL
ace011	Use of Low-Cetane Fuel to Enable Low-Temperature Combustion	ANL
ace024	Particulate Emissions Control by Advanced Filtration Systems for GDI Engines	ANL
ace054	RCM Studies to Enable Gasoline-Relevant Low-Temperature Combustion	ANL
ace075	Advancements in Fuel Spray and Combustion Modeling with High-Performance Computing Resources	ANL
ace084	High-Efficiency GDI Engine Research, with Emphasis on Ignition Systems	ANL
ace061	ATP-LD; Cummins Next-Generation Tier 2 Bin 2 Diesel Engine (Projected engine in 2015)	Cummins
ace094	Ultra-Efficient Light-Duty Powertrain with Gasoline Low-Temperature Combustion	Delphi Powertrain
ace092	High-Efficiency VCR Engine with Variable Valve Actuation and New Supercharging Technology	Envera LLC
ace089	Development of Radio Frequency Diesel Particulate Filter Sensor and Controls for Advanced Low-Pressure Drop Systems to Reduce Engine Fuel Consumption	Filter Sensing Technologies, Inc.
ace065	Advanced Gasoline Turbocharged Direct Injection GTDI Engine Development (Project ended in 2015)	Ford Motor Company
ace093	Lean Miller Cycle System Development for Light-Duty Vehicles	General Motors
ace012	Model Development and Analysis of Clean and Efficient Engine Combustion	LLNL
ace013	Chemical Kinetic Models for Advanced Engine Combustion	LLNL
ace076	Improved Solvers for Advanced Engine Combustion Simulation	LLNL
ace014	2015 KIVA-hpFE Development: A Robust and Accurate Engine Modeling Software	LANL
ace079	Robust Nitrogen Oxide/Ammonia Sensors for Vehicle On-board Emissions Control (Project engine in 2015)	LANL
ace087	Next-Gen Ultra-Lean Burn Powertrain (Ended in 2005)	MAHLE Powertrain LLC
ace015	Stretch Efficiency for Combustion Engines: Exploiting New Combustion Regimes	ORNL

continued

TABLE 3-2 Continued

Project ID	Presentation Title	Organization
ace016	High-Efficiency Clean Combustion in Multi-Cylinder Light-Duty Engines	ORNL
ace017	Accelerating Predictive Simulation of Internal Combustion Engines with High-Performance Computing	ORNL
ace022	Joint Development and Coordination of Emissions Control Data and Models CLEERS Analysis and Coordination	ORNL
ace032	Cummins/Oak Ridge National Laboratory-FEERC CRADA: NOx Control and Measurement Technology for Heavy-Duty Diesel Engines	ORNL
ace033	Emissions Control for Lean Gasoline Engines	ORNL
ace052	Neutron Imaging of Advanced Transportation Technologies	ORNL
ace077	Cummins/Oak Ridge National Laboratory-FEERC Combustion CRADA: Characterization and Reduction of Combustion Variations	ORNL
ace085	Low-Temperature Emission Control to Enable Fuel-Efficient Engine Commercialization	ORNL
ace090	High-Dilution Stoichiometric Gasoline Direct-Injection SGDI Combustion Control Development	ORNL
ace023	CLEERS: Aftertreatment Modeling and Analysis	PNNL
ace026	Enhanced High- and Low-Temperature Performance of NOx Reduction Materials (Projected engine 2015)	PNNL
ace027	Thermally Stable Ultra Low-Temperature Oxidation Catalysts	PNNL
ace056	Fuel-Neutral Studies of Particulate Matter Transport Emissions	PNNL
ace078	Investigation of Mixed Oxide Catalysts for NO Oxidation	PNNL
ace091	Intake Air Oxygen Sensor	Robert Bosch
ace002	Light-Duty Diesel Combustion	SNL
ace004	Low-Temperature Gasoline Combustion LTGC Engine Research	SNL
ace005	Spray Combustion Cross-Cut Engine Research	SNL
ace006	Automotive Low-Temperature Gasoline Combustion Engine Research	SNL
ace007	Large Eddy Simulation LES Applied to Advanced Engine Combustion Research	SNL
ace095	Metal Oxide Nano-Array Catalysts for Low-Temperature Diesel Oxidation	University of Connecticut

NOTE: Green, premixed flame dominated (SI); orange, mixing/diffusion dominated (CI); red, low-temperature combustion; blue, CFD development; yellow, aftertreatment; multiple colors, a project indicates it contributes to multiple areas. Acronyms defined in Appendix D.
SOURCE: Project numbers are from the Department of Energy's 2016 Annual Merit Review, see Cooper (2016a).

Premixed Flame Dominated (SI) Engines—Dilute Gasoline

Premixed flame dominated, dilute combustion engines are currently constrained by knock-limited operating regimes, lack of combustion robustness, low combustion rates, low exhaust enthalpy, gas exchange complexity (such as EGR and air handling), and high emissions of hydrocarbons (HCs) and NO_x. Lean or dilute combustion offers higher efficiency; however, as one pushes the limits of dilution it is more difficult to initiate and maintain the flame. Once the flame is established the traditional approach of enhancing the burning velocity by increasing turbulence is limited, because it is easier to extinguish the flame under the higher fluid shear that occurs when turbulence is increased. If the dilution is obtained through EGR, either internal or external, the lack of homogeneous composition of the EGR with the intake air can cause flame extinction much more easily than for stoichiometric flames. Similarly, the flame propagation is more sensitive to mixture inhomogeneous conditions that occur between the fuel and the air in the combustion chamber under conditions of high dilution. Cycle-by-cycle variations during the engine operation now become a more serious problem. Because the combustion is less stable and more easily extinguished, unburned hydrocarbons and carbon monoxide (CO) emissions can become excessive. As boost is used to increase the power density of the engine, knock limits are often reached during high load operation. This limits the potential improvement obtainable from the engine because actions taken to avoid knocking operation compromise performance. Finally, when operating under very dilute conditions the exhaust temperature becomes so low that it taxes the capability of the after-treatment system. If the high dilution is achieved through lean operation, three-way catalysts are no longer effective for NO_x control, so lean NO_x reduction would need to be implemented. Projects that contribute to further understanding of these phenomena are color coded green in Table 3-2.

Mixing and Diffusion-Dominated (CI)—"Clean Diesel"

Diesel engines have a potential efficiency advantage over premixed flame-dominated engines but they also have significant challenges. The ACECTT has identified the most critical challenges being faced by engineers working on diesel engines as follows: in-cylinder NO_x and soot control, EGR and air handling, fuel injection and control systems, and the cost and complexity of the systems. The characteristics of the fuel injection process play an important role in achieving clean diesel combustion. The intricate details of the internal geometry of the injector and how it affects the fuel flow leaving the injector play an important role in the early entrainment and mixing of the fuel and air in the immediate vicinity of the injector tip. This also impacts the symmetry of the spray pattern among the individual injector nozzle-hole plumes. The early behavior of the individual fuel plumes leaving the nozzle determine the air-to-fuel ratio of the portion of the spray that is first to experience autoignition, which in turn has a high impact on the initial particulate formation rate. The formation of the particulate at this early part of the combustion process subsequently dictates the characteristics of the soot-NO_x trade-off of the engine. EGR is an important aspect of the soot-NO_x trade-off of the engine. Since diesel engines typically do not have throttles in the intake manifold, the pressure difference between the intake and the exhaust manifold is minimal. Engine performance is easily compromised by efforts made to induce the exhaust gas to flow between the manifolds when there are either insufficient or adverse pressure gradients. Different approaches to getting the exhaust gas from the exhaust to the intake manifold, like high-pressure and low-pressure loops, along with the challenges this imposes on the operation of the boosting system and the resulting impact on the pumping work, are motivating much technical research and development. Projects listed in Table 3-2 addressing these fundamental challenges are color coded orange.

Low-Temperature Combustion

Engines operating using LTC processes have the potential for very high efficiency with low cylinder-out emissions. However, there are significant challenges to implementing this combustion mode in an engine. Chief among them are high combustion noise, achieving combustion robustness, high engine-out HCs and CO, transient combustion control and emissions, cold start ability, obtaining a wide speed and load operating range, and cost and complexity. The combustion initiation for LTC operation depends on the chemical kinetic autoignition of a partially premixed air-fuel mixture. There needs to be a certain degree of nonuniformity within the air-fuel mixture in the cylinder to obtain acceptable combustion. If the mixture is too uniform the entire mixture autoignites at once and the rate of combustion is excessive. If the mixture is too stratified the combustion can either become excessively long or fail to go to completion. The nonuniformity of the mixture can be manipulated in many ways: nonuniformity in temperature, in air-fuel ratio, in oxygen concentration, or in fuel reactivity. In addition, the level of the nonuniformities in the cylinder necessary for good combustion is intimately linked to the characteristics of the fuel. This makes control of all aspects of the in-cylinder conditions extremely important, so engine controls and transients are a major technical challenge. Even if LTC modes were used in hybrid applications there would still be transients because of the stop-start nature of the engine operation in hybrid vehicles. Projects that contribute to further understanding of these phenomena are color coded red in Table 3-2.

Advancements in Computational Fluid Dynamic Capabilities

An important component of all of the ACECTT research effort is advanced computer simulation. New understandings from research are integrated into the simulations, which in turn are used to offer higher-fidelity interpretations of the experimental results and to guide further research activities through prediction of more optimal operating conditions. In the quest of ever better engine performance the need for deeper understanding of the governing fundamentals continues to grow, which subsequently requires higher resolution and more precise submodels to be integrated into the simulations. For example, kinetic models capable of predicting the impact of different fuel compositions on engine performance; higher-fidelity numerical submodels such as fluid turbulence models simulating mixing and dissipation to ever smaller physical scales; combustion models that can capture the transition between flame propagation, mixing-controlled burning, and LTC energy release; better submodels for the heat transfer processes in the cylinder; more accurate emission models; more robust and faster numerical algorithms; and increased capability for program parallelization are examples of important advancements that would make the simulation efforts even more valuable than they are now. Projects listed in Table 3-2 addressing these fundamental challenges are color coded blue.

Aftertreatment Systems Advancements

As the engines are made more efficient the energy in the exhaust gases leaving the engine decreases, which requires the aftertreatment systems to function at lower temperatures. Furthermore, as the number of people living in urban areas increases, vehicle density in those areas increases. This results in a larger input of engine exhaust into the urban environment, which drives the need for even lower regulatory limits on criteria pollutants. Consequently, the exhaust gas aftertreatment systems need to be made more effective and capable of functioning at lower temperatures than current systems. This establishes the research priorities for exhaust gas aftertreatment systems. The aftertreatment systems are integral to the engine and power train; they need to be effective at temperatures below the current operating temperatures of 400°C to 600°C, reach light-off temperatures as quickly as possible, be resilient to poisoning from contamination, use minimal precious metals, and effectively filter particulate matter down to a size of 23 nm, the particle diameter above which the European number regulations are enforced, and they need to be easily regenerated. Projects that contribute to further understanding of these phenomena are color coded yellow in Table 3-2.

The deeper understanding of the thermodynamics, fluid mechanics, heat transfer, combustion physics, fuel chemistry, and catalysis that is sought in the research projects in Table 3-2 is necessary to develop better engine components and exhaust gas aftertreatment systems, but it is only part of the solution to achieve maximum performance with minimal environmental impact from the vehicle. The engine, exhaust gas aftertreatment components, and the power train must work as an optimized system, which presents complex challenges in total system control. These challenges not only push current capabilities in control theory but also require advancement in the current state of the art of sensors, actuator technology, and onboard computer power.

The projects that the ACECTT advocates supporting strive to enhance the analysis capabilities and fundamental understanding of the technical barriers to more efficient and environmentally benign engines. As can be seen from the list of projects with which the U.S. DRIVE ACECTT engages (Table 3-2), research activities encompass a wide range of fundamental phenomena, which can be summarized topically as follows:

- Measuring and simulating phenomena occurring during the fuel injection process,
- Developing increased understanding and appropriate models for the combustion kinetics of different combustion strategies and their associated emissions,
- Developing sensors and new measurement techniques,
- Working on controlling combustion robustness and repeatability,
- Developing higher-fidelity turbulence models,
- Developing predictive simulations with high-performance computing,
- Developing combustion control algorithms,
- Working on aftertreatment modeling and analysis,
- Trying to enhance low-temperature catalysis, and
- Developing the next generation of an advanced modular computer program, which will facilitate inclusion of the new and higher-fidelity subprograms necessary to integrate predictive simulation into the design process.

In addition to the fundamental research activities, the ACECTT also engages in industry cost-shared demonstration programs. The purpose of these programs is to push the envelope of current product performance by integrating suites of advanced technologies from the research laboratory into a power train system and/or a vehicle. In doing this, valuable learning occurs as to the subtleties of interactions between the different system components and power train controls once they are completely integrated into a product. It also helps guide the manufacturers as to the time frame when different advanced technologies might be commercially viable. Within U.S. DRIVE four such activities have been identified:

1. ATP-LD, Cummins Next-Generation Tier 2 Bin 2 Diesel Engine;
2. Ultra-Efficient Light-Duty Power Train with Gasoline Low-Temperature Combustion, with Delphi;
3. Advanced Gasoline Turbocharged Direct Injection (GTDI) Engine Development, with Ford; and
4. Lean Miller Cycle System Development for Light-Duty Vehicles, with GM.

Overall the ACECTT is engaged in a broad array of research projects, ranging from fundamental laboratory investigations, to understanding subtle phenomena occurring within commercially available technologies, to exercises in advanced technology integration for market readiness assessment. With such a broad range of research activities it is important to make sure that efforts are synergistic and not duplicative, and that time horizons and risk factors of the research are appropriate to the organization performing the work. The advanced technology integration projects shown earlier can be very informative and are not something that industry would do when high-risk unproven technologies are being used. The national laboratories are not involved in these activities, which is appropriate. Within the ACECTT programs some of the research within the national laboratories involves use of already developed engine technologies. It is important that such work address underlying fundamentals that need to be better understood, and if understood would open up new performance possibilities. It should be work that industry would not or could not do on its own and that benefits the technical community at large. It appears to the committee that the ACECTT is doing a good job at maintaining this focus within their diverse program and that the DOE AMR serves as a good venue for getting feedback from the technical community.

Current Status vis-à-vis Goals and Targets

The principal barrier to achieving U.S. DRIVE's research targets is lack of fundamental understanding of the relevant physics, chemistry, and thermodynamics. Consequently, the technical team's status and progress toward reaching its goals is best measured by advances in knowledge, the effectiveness of the communication of that knowledge to the appropriate stakeholders, and ultimately the application of that knowledge to achieve an improvement in an engine-power train system.

The ACECTT has established a series of networks to facilitate communication and knowledge transfer within the appropriate technical communities. By interfacing with USCAR there is regular information exchange on fundamental needs as well as feedback on the programs considered to be part of the U.S. DRIVE portfolio. For example, industry representatives interface through USCAR to give critiques of the U.S. DRIVE projects presented at the DOE AMR each year. These critiques are in addition to the evaluations performed by the designated reviewers at the AMR meeting and are an important part of a checks and balances system of assessing if the work being performed by the different groups within the Partnership is appropriate to that group's capability and time horizon.

Close collaboration with industry is facilitated through the Advanced Engine Combustion Memorandum of Understanding (MOU) led by Sandia National Laboratories. The MOU works to carry national laboratory research into products via crosscutting efforts between light-duty and heavy-duty engine R&D. Almost all major engine manufacturers and energy companies participate in the MOU. University research is also integrated with the MOU.

The Engine Combustion Network is an excellent example of a predominantly web-based facilitation of data and CFD analysis exchange for better computer models of spray combustion. Over 16 international teams share experimental data, CFD approaches, codes, and actual model submissions for a well-characterized set of injectors operating under accurately prescribed conditions. This allows DOE/VTO investments to be leveraged many times over by complementary experiments conducted by other institutions. The collaboration efficiently transfers the research results to industry's workflow through the development of engineering models. U.S. DRIVE estimates that the combined effort has accomplished 15 years of research progress in approximately 3 years (Howell, 2016).

As noted in the NRC Phase 4 report (NRC, 2013), the Cross Cut Lean Exhaust Emissions Reduction Simulation (CLEERS) is a technology focus group whose purpose is to promote development of improved computational tools for simulating realistic full-system performance of lean-burn diesel or gasoline engines and associated emissions control systems. CLEERS holds annual meetings as well as monthly phone conferences. Figure 3-2 shows the structure of the CLEERS focus group. It has proven to be a very effective venue for technical exchange between broad scopes of participants. The structure has proven so successful that the DOE Bioenergy Technologies Office (BETO) has adopted the model for their Computational Pyrolysis Consortium.

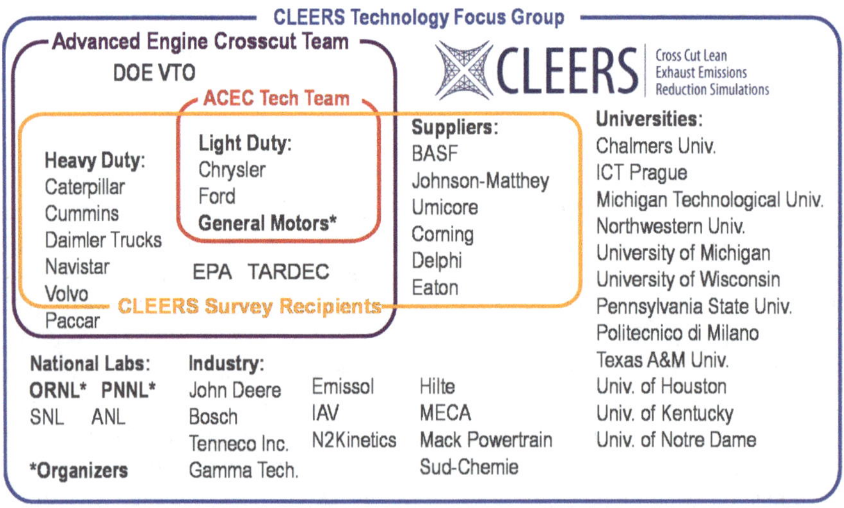

FIGURE 3-2 Schematic of the organizational structure of CLEERS (Cross Cut Lean Exhaust Emissions Reduction Simulations). SOURCE: Solomon and Howden (2016).

In addition to the preceding structured programs, which have regularly scheduled interactions, the ACECTT has established collaboration through special workshops, such as the following:

- Future Automotive Aftertreatment Solutions: The 150°C Challenge Workshop (2012);
- Low-Temperature Oxidation Catalyst Test Protocol (2014);
- CFD for Engines Workshop (2014);
- Recommendations for Future Fuel Properties (2015); and
- Low-Temperature Combustion (LTC) Fuel Properties Survey (work in progress).

The ACECTT is actively engaged in following the research activities within the DOE VTO portfolio. Through the collaborations, working groups, workshops, and review participation they stay informed of the critical research needs of the industry partners and are effectively facilitating the knowledge exchange that supports the industry goals of more efficient, cleaner vehicles.

The assessment of the committee is that the U.S. DRIVE ACECTT is doing an excellent job at advancing the technical community's understanding of the fundamental barriers limiting the improvement of engine efficiency and cleanliness and in disseminating that understanding to the relevant stakeholders.

Assessment of Progress and Key Achievements

When asked by the committee, the ACECTT submitted the following list of accomplishments for their combustion and aftertreatment activities:

- *Combustion*
  - —Cummins ATLAS Diesel Engine Program Achieves 60% Cycle Fuel Economy Improvement over Gasoline Baseline—Cummins (2015);
  - —Lean Downsized Boosted Engine Projected to Improve Combined-cycle Fuel Economy by 25%—GM (2013);
  - —Gasoline Direct-Injection Compression Ignition Shows Potential for 39% Fuel Economy Improvement—Delphi (2014);
  - —New Method Developed to Assess Engine Efficiency Opportunities—Oak Ridge National Laboratory (ORNL) (2013);
  - —Unique Spray Measurements Enable Improved Models—Sandia National Laboratories (SNL) and Argonne National Laboratory (ANL);
  - —Experimentally Validated High-Fidelity Simulations Enable Improved Gasoline Direct Injection Fuel Injection System Designs—ANL (2015); and
  - —First Principles Simulations Reveal Details of Fuel Injection that Control Combustion and Emissions—SNL (2015).

- *Aftertreatment*
  - —Study of GDI PM Size, Shape, and Composition Highlights Control Challenges That Differ from Diesels—ORNL (2013);
  - —Low-Temperature Oxidation Catalyst Test Protocol—Advanced Combustion and Emission Control (2014);
  - —Innovative Metal Oxide Catalyst Improves Low-Temperature Reactivity of Traditional Oxidation Catalyst—ORNL (2015); and
  - —Promising Automotive Catalyst Provides Nitrogen Oxide Control at Temperatures Approaching 150°C—Pacific Northwest National Laboratory (PNNL) (2015).

Significant Barriers and Issues That Need to Be Addressed

The barrier to further improvement in engine efficiency and more effective aftertreatment systems continues to be the incomplete understanding of the detailed fundamentals of the thermodynamic, fluid mechanic, heat transfer, chemical kinetic, and combustion physics processes that occur within the engine and the aftertreatment systems. The tremendous advancements in engineering capabilities that allow designing and manufacturing with specifications on a molecular level gives confidence that as the engine community's understanding becomes more complete, additional improvements in engine efficiency and cleanliness can be made.

Responses to Recommendations from the Phase 4 Report

NRC Phase 4 Review Recommendation 3-1. The DOE should undertake a larger effort on the next generation of KIVA in order to be successful in facilitating such a resource. There should be a more formal collaboration established among the industry stakeholders, university stakeholders, and the DOE researchers doing the development work for KIVA IV.[2] Efforts should be made to implement a modular and object-oriented structure to the code that is most useful to the ultimate stakeholders.

Partnership Response. DOE's Vehicle Technologies Office (VTO) agrees that a larger effort is needed to develop the next generation of KIVA. Based on availability of funds, VTO may increase resources devoted to KIVA. The Advanced Engine Combustion Memorandum of Understanding (MOU) facilitates collaboration among numerous stakeholders. MOU members include BP, Caterpillar, Chevron, Chrysler, Cummins, Detroit Diesel, ExxonMobil, Ford, General Electric, General Motors, John Deere, Navistar, Phillips 66, Shell, and Volvo. A more formal collaboration exists between the DOE researchers at Los Alamos National Laboratory (LANL) and several universities that are subcontractors to LANL. Also, several companies have inquired about starting more formal collaborations with LANL. LANL has a modular and object-oriented engine modeling code in development that will be more predictive and much more robust, while allowing for very rapid grid generation without compromising accuracy.

[2] KIVA is the name of a 3-D CFD program that is being developed by DOE that can be used for engine combustion studies.

Committee Assessment of Response to 3-1. The committee thanks U.S. DRIVE for their efforts to improve the utility of the CFD modeling effort. The committee is aware that DOE held a high-performance computing workshop in August 2014 with invitations to all stakeholders. The outcomes of the workshop were a prioritized list of attributes heralded by industry as necessities, and plans for commercialization of the code DOE is developing with a goal of maintaining mostly open-source source code so university students can continue to modify and gain familiarity with the code for their own purposes and training. These outcomes address the concerns raised by the committee in Recommendation 3-1 of the Phase 4 review. The committee endorses the current program path and the continued development of KIVA-hpFE.

Appropriateness of Federal Role

The committee believes it is appropriate for the federal government to maintain a role in combustion and emissions control research. As described in the introduction to this section, there is still significant opportunity to reduce the fuel consumption and environmental impact of ICE-powered vehicles, so it is important to keep an active research program in this area. Developing the enhanced understanding and tools to do this pushes the state of the art in almost all physical and engineering sciences.

Findings and Recommendations

Finding 3-1. The ACECTT is a well-organized group within the U.S. DRIVE organization. They understand the technical barriers that need to be overcome to further increase the efficiency and environmental friendliness of ICE-powered vehicle systems.

Finding 3-2. To guide the technical work toward overcoming technical barriers to higher efficiency, the advanced combustion and emissions control technical team (ACECTT) has established stretch efficiencies goals for 2020 for peak and intermediate engine loads for the three types of engine power train systems they expect to be most prevalent in the near term: hybrid applications, naturally aspirated, and downsized boosted engine systems. The ACECTT is also engaged in research activities in chemical kinetic development, and promoting a more fundamental understanding of the interaction between fuel characteristics—such as Research and Motor Octane number, heat of vaporization, etc.—and different engine operating conditions. This work is aimed at facilitating the integration of advanced kinetically controlled combustion processes, i.e., low-temperature combustion, as part of the engines' operating map, which is considered a longer-term technology.

Finding 3-3. The ACECTT focus for both near- and longer-term research is centered on conventional four-stroke engine architectures. However, work on alternative engine architectures is taking place. Some of that work is under DOE funding, and claims are being made in the literature of potential efficiency and environmental impact improvements for these different engine architectures (Redon et al., 2014; Chadwell et al., 2014).

Recommendation 3-1. The advanced combustion and emissions control technical team should be proactive in seeking out and assessing data on the performance of alternative engine architectures and concepts that will allow benchmarking against those within their current research portfolio.

Finding 3-4. The ACECTT has recognized that fuel characteristics can be a contributor to improving engine efficiency and reducing environmental impact. Because of the broadening of fuel feedstocks to nontraditional gas and oil as well as biomass, it is appropriate to include fuel characteristic considerations into ACECTT's combustion research portfolio.

Finding 3-5. A Fuels Working Group has been formed and is actively engaged with the ACECTT, USCAR, and the DOE multilaboratory initiative Co-optimization of Fuels and Engines.

Finding 3-6. The ACECTT is engaged in a broad array of research projects ranging from fundamental laboratory investigations, to understanding subtle phenomena occurring within commercially available technologies, to exercises in advanced technology integration for market readiness assessment. There appears to be good collaborative engagement among the industry stakeholders (engine and vehicle manufactures, Tier 1 suppliers, and energy companies), USCAR, and the ACECTT. The committee believes that the ACECTT is doing a good job making sure the individual efforts are synergistic and not duplicative and that the time horizons and risk factors of the research are appropriate to the organization performing the work.

COMBUSTION ENGINE FUEL AND LUBRICANT TECHNOLOGIES

Petroleum

Background

The U.S. petroleum industry has evolved significantly since the NRC Phase 4 review of U.S. DRIVE (NRC, 2013). The development of hydraulic fracturing ("fracking") and directional drilling techniques for production of both oil and natural gas has increased domestic energy supplies (EIA, 2016a), and reduced energy imports to the point where it is predicted that the United States will be a net exporter of energy by 2040. Per the Energy Information Administration, "By 2040, total U.S. energy production is greater than total U.S. energy consumption, allowing for U.S. net energy exports equal to 4% of total consumption" (EIA, 2016a). As a result, the development of these alternative production methods has ensured the use of petroleum as one of the primary transportation fuels for light-duty vehicle applications for decades to come.

DOE and U.S. DRIVE interest in transportation fuel research is strongly influenced by three U.S. regulatory initiatives:

- EPA Tier 3 Emissions Regulations,
- Fleet CAFE [corporate average fuel economy] Standards for Light-Duty Vehicles of 54.5 mpg by 2025, and
- Renewable Fuel Standard for use of 36 billion gallons of renewable fuels (other than corn ethanol) by 2022.

The potential for ample amounts of petroleum-based fuel for use in transportation has increased concerns regarding the issue of GHG emissions and their effects on climate change. For that reason, it is critical that engine design and fuel composition be considered a system in developing new technologies for improving vehicle fuel efficiency and reducing CO_2 emissions. To this end, as described earlier in this chapter, the U.S. DRIVE ACECTT has included light-duty automotive fuel research projects within the mission of their research and development program (U.S. DRIVE, 2013h). This R&D effort has two overall goals:

1. To reduce our nation's dependence on petroleum for transportation by conducting R&D to enhance the use of "drop-in" fuels from alternative sources,[3] especially low-carbon fuel sources, and
2. To determine fuel characteristics that enable current and emerging advanced combustion engines and aftertreatment systems that meet program [U.S. DRIVE] objectives.

These goals require identifying how drop-in fuels will impact advanced combustion and emissions control strategies as well as identifying practical, economic fuels and fuel-blending components with potential to directly replace significant amounts of petroleum (U.S. DRIVE, 2013h).

Parasitic losses in the engine and other driveline components are also a source of wasted fuel energy. Fenske et al. (2015) have calculated that for the 250 million "on-road" vehicles in the United States the daily loss in fuel energy due to friction is 1.5 to 1.8 million barrels of fuel. To address this issue, DOE is also funding research on advanced, low-friction lubricant formulations in support of U.S. DRIVE goals and objectives.

Petroleum fuel and lubricant research in support of the U.S. DRIVE goals is budgeted through the DOE VTO. Currently there are eight projects being con-

[3] Fuels from alternative sources that are chemically equivalent to petroleum fuels.

ducted by DOE laboratories and supplier partners that specifically address fuel and lubricant research efforts. However, as described earlier in this chapter, there are additional advanced combustion projects that also include the evaluation of fuel formulations in meeting engine efficiency and emissions goals.

Table 3-3 provides an overview of DOE's VTO projects specifically related to petroleum-based fuels, petroleum/bio-based fuels, and lubricant research activities.

The total DOE budget for VTO fuel and lubricant subprograms in 2014 was $15.5 million and in 2015 was $20.0 million, for a total budget over the previous 2 years of $35.5 million. A total fuel and lubricant budget of $37 million has been requested for 2016. These projects are distributed over a number of national laboratories, original equipment manufacturers (OEMs), and numerous suppliers. In the case of fuel research, the projects are focused on not only the combustion and spray characteristics of petroleum-based fuels (both gasoline and diesel) but also the characteristics of petroleum-biofuel blends. The use of biomass or nontraditional petroleum as either feedstocks or drop-in components to the fuel not only reduces the fuels' carbon footprint, but it also introduces the possibility of changing the fuel characteristics in order to optimize the fuel and the engine together.

TABLE 3-3 Department of Energy Vehicle Technologies Office U.S. DRIVE-Related Fuel and Lubricant Projects List

Project ID	Project Title	Organization
ft029	Additive and Basefluid Development	Argonne National Laboratory
ft025	Improve Fuel Economy through Formulation Design and Modeling	Ashland
ft026	Developing Kinetic Mechanisms for New Fuels and Biofuels	Lawrence Livermore National Laboratory
ft002	Advanced Combustion and Fuels	National Renewable Energy Laboratory
ft007	Fuel Effects on Emissions Control Technologies	Oak Ridge National Laboratory
ft008	Gasoline-Like Fuel Effects on Advanced Combustion Regimes	Oak Ridge National Laboratory
ft004	Fuel Effects on Mixing-Controlled Combustion Strategies for High-Efficiency Clean-Combustion Engines	Sandia National Laboratories
ft006	Advanced Lean-Burn DI Spark Ignition Fuels Research	Sandia National Laboratories

NOTE: DI, direct injection.
SOURCE: Project numbers are from the Department of Energy's 2016 Annual Merit Review, see Cooper (2016a).

Experimental data collected from individual projects are used to develop injector spray flow patterns and kinetic models of fuel combustion that are subsequently used in large CFD calculations of in-cylinder combustion. These programs are important in providing a methodology for efficiently evaluating a large number of potential options for fuel composition, injection strategies, and cylinder designs. The national laboratories have been able to develop new analytical techniques for identifying the combustion properties of fuel blends (e.g., Knock Index) in laboratory experiments using minute amounts of fluid, as well as new optical techniques for analyzing the formation of soot within an engine cylinder. In most cases the analytical facilities of the national laboratories greatly exceed the facilities at supplier and OEM laboratories and for this reason the use of the national laboratories for this work is appropriate in meeting U.S. vehicle fuel efficiency and emissions goals.

Status of U.S. DRIVE Fuel and Lubricant Goals and Targets

In support of the U.S. DRIVE goals for fuel and lubricant research, the VTO has directed a portion of the R&D they support toward achieving specific light-duty vehicle fuel and lubricant targets. As described to the committee (Howell, 2016), these targets include the following:

- *Fuel Targets*
  - —Identify a spark-ignition candidate fuel "downselect" in 2017 (available at large scale, early 2020s),
  - —Complete R&D on advanced conventional fuel/engine systems meeting 2025 CAFE/GHG target in 2020, and
  - —Demonstrate optimized kinetically controlled engine/fuel system providing 30 percent GHG reduction versus "business as usual" in 2025
- *Lubricant Target*
  - —Demonstrate an engine/driveline lubricant package providing 4 percent fuel economy improvement relative to 2010 base fluids in 2020.

In order to achieve these goals, the VTO has funded the specific projects listed in Table 3-3. Although most of these projects are managed by individual national laboratories, the project teams include significant participation by U.S. DRIVE OEMs as well as other auto industry partners and suppliers. In addition, DOE has also funded supporting research at academic laboratories throughout the United States. For example, DOE recently announced a $7 million solicitation for research on advanced fuels for use in high-efficiency, low-emissions ICEs (DOE, 2016d). This solicitation was restricted to U.S. institutes of higher education. The final selection of academic research teams for this solicitation involves researchers from 14 universities. Using teams made up of governmental, industrial, and academic researchers, this research is intended to promote the rapid commercial introduction of viable technologies developed by industry using the DOE research project results.

As described earlier in this chapter, U.S. DRIVE has recently established a "crosscut" activity titled the Fuels Working Group that includes not only U.S. DRIVE members and national laboratory personnel but also one non-U.S. DRIVE fuel company. Within the U.S. DRIVE organization, "working groups" are a less formal collaborative activity than are technical teams. Working groups do not conduct research according to a roadmap, nor do they have specific technical targets. Instead, they support the goals of related technical teams by coordinating research projects among interested parties both within and beyond U.S. DRIVE. In this regard, the FWG maintains close technical cooperation with the ACECTT. The specific mission of the FWG is to "evaluate potential properties of lower carbon fuels for future, high efficiency engines and combustion regimes working in close coordination with the advanced combustion and emission control tech team" (Cooper, 2016b).

To meet this mission, the FWG has identified in cooperation with the ACECTT four focus areas for their research programs. These focus areas are designed to evaluate potential properties of lower carbon fuels for future, high-efficiency engines and combustion regimes meeting ACECTT targets. These focus areas are as follows (Farenback-Brateman, et al., 2016):

1. Premixed, Flame Propagation, Spark Ignition Combustion Mode (SI)
2. Mixing/Diffusion Compression Ignition Combustion Mode (CI)
3. Chemical Kinetics Dominated Low-Temperature Combustion Modes (LTC)
4. New Combustion Quality Metrics
   - Anti-Knock for SI
   - Ignition Delay for LTC

As an example of how the FWG will address these focus areas, the ACECTT has identified a set of specifications for an advanced gasoline formulation shown in Table 3-4, which it believes would greatly enhance spark-ignition engine performance due to allowing higher compression ratios and reducing the need for spark retardation. To determine if these specifications will in fact meet the purposes identified in Table 3-4, the FWG is actively involved, as part of their focus area 1, in collecting engine data intended to define the benefits of each specification. These engine tests are being conducted within OEM laboratories and national laboratories.

In addition, the data collection efforts are being supported by several research projects being conducted within the Coordinating Research Council (CRC). The CRC is an independent research organization that is made up of OEM and oil company members. Recently, national laboratory personnel have also joined CRC working groups and contributed significantly to the identification of sur-

TABLE 3-4 Summary of Advanced Combustion and Emission Control Technical Team (ACECTT) Recommendations for Future Spark-Ignition Fuels

Property	Recommended Target Value	Purpose
Research Octane Number	>100	Engine efficiency
Octane sensitivity S = (RON – MON)	S > 12 (or MON < 88)	Engine efficiency
Sulfur	10 ppm max	Emissions control
Volatility	Reduced variability in Driveability Index	Emissions control
Properties governing particulate matter	Particulate Matter Index < 1.5	Emissions control
Heat of Vaporization (HoV)	HoV > Current gasolines	Engine efficiency

NOTE: MON, motor octane number; RON, research octane number.
SOURCE: Solomon and Howden (2016).

rogate fuel compositions that can be used to simulate commercial fuel combustion in advanced engines. The engine and fuel data collected from all of these sources will be used to model vehicle-level CO_2 emissions and fuel economy using "Autonomie Software" (Farenback-Brateman et al., 2016). At the same time Argonne National Laboratory will conduct "well-to-wheels" calculations of combined fuel production and vehicle efficiency using the GREET model. The data from all of these sources will be used to calculate a C2G analysis of the CO_2 footprint for each engine/fuel combination. The FWG will also identify and document infrastructure implications of new fuel formulations and identify technical roadblocks to implementation. All of this effort is directly related to determining the efficacy and benefits of the FWG focus area 1. Similar efforts and research projects will also be developed to quantify the benefits of the other three FWG focus areas.

As described earlier in this chapter, the FWG is also coordinating its efforts with the consortium within Energy Efficiency and Renewable Energy (EERE) called "Co-Optima" (Howell, 2016; Sarkar, 2016; Farenback-Brateman et al., 2016; Farrell, 2016). The Co-Optima initiative involves an External Advisory Board including not only membership of USCAR and the Truck and Engine Manufacturers Association but also membership of the American Petroleum Institute, the Advanced Biofuels Association, the EPA, the California Air Resources Board, and other industry groups. Current fuels are predominantly petroleum based, and refinery systems have been optimized for producing the current mix of gasoline and diesel. Their respective properties have traditionally been defined by meeting regulatory requirements and agreements reached within consensus organizations like the American Society for Testing and Materials (ASTM) and between members of the vehicle and energy industries. However, from the perspective of minimizing GHG emissions within the context of more advanced engine technologies with a broader base of fuel feedstocks, Co-Optima will evaluate nontraditional sources of petroleum and biomass-derived supplies to determine if it is possible to specify fuel characteristics that support more efficient engine operation. In addition, it is important to determine if optimization of fuel properties could further facilitate the introduction of advanced combustion processes, such as kinetically controlled, low-temperature combustion. However, it is important to note that identifying and ultimately implementing newly optimized fuels in the commercial marketplace is a huge task because any transition in fuel characteristics needs to be done while (1) maintaining compatibility with current infrastructure and legacy fleets, and (2) meeting current EPA/NHTSA fuel economy and GHG regulations as well as individual state fuel specifications based on industry consensus agreements such as those developed within the ASTM.

A recent status report on the developments generated within the first year of the Co-Optima consortium has been published (NREL, 2017). The introduction to this status report suggests that the objective to Co-Optima is

> To arm industry, policymakers, and other key stakeholders with the scientific foundation and market intelligence required to make investment decisions, break down barriers to commercialization, and bring new high-performance fuels and advanced engine systems to market sooner.

The status report summarizes recent research results that will significantly benefit USCAR OEMs in the development of advanced fuel-efficient, low-GHG engines. The status report maintains that the consortium will enable the introduction of new commercial fuel and engine technologies by 2025. As described in the following material, the committee understands that the Co-Optima initiative is a major undertaking and that meeting its stated goals, although well intended, will be difficult given the stated timeline.

Although the Co-Optima consortium is not a specific activity within U.S. DRIVE and therefore not a subject of this review, the existence of Co-Optima and its inclusion of many additional stakeholders with diverse interests raises several questions of importance to U.S. DRIVE such as how the initiative will be conducted, how decisions on experimental fuel properties and compositions will be made, and how advanced fuel specifications will be used. For example, if a set of fuel properties is identified to be useful in increasing engine efficiency or reducing C2G CO_2 emissions, will these properties be allowed or recommended for use in

- A laboratory fuel for advanced engine tests to demonstrate the benefits of co-optimized engine/fuel systems,
- A certification fuel that could be used by manufacturers in emissions testing of vehicles containing advanced engine technologies, or
- A commercial fuel available in the marketplace for general customer use?

There is also a question of how conflicts of opinions between the stakeholders directed at a common goal—engine/fuel combinations optimized for low emissions and high efficiency—will be adjudicated and resolved.

In addition, it is important to recognize that even if Co-Optima identifies an "optimum" engine/fuel combination for improving fuel efficiency and reducing GHG emissions, OEMs will require an extended amount of time for evaluating and validating the performance of such a fuel in their advanced development engines. Fuel effects on the durability of the fuel tank, pump, and injector systems will need to be verified. Fuel effects on the generation of criteria emissions including "toxics" and evaporative emissions will have to be determined over extended mileages. The fact that advanced engine/fuel combinations meet efficiency and emissions targets in national laboratory tests does not mean that OEMs will immediately adopt this proposed optimum for use in commercial vehicles. Correspondingly, energy companies will not introduce new fuels into the commercial marketplace without confidence that the engine systems that rely on such fuels will be available. Each of these issues will slow the introduction of new engine/fuel concepts despite the value of the advanced research being conducted and the new concepts being generated.

Progress and Key Achievements

There has been significant progress in meeting both fuel and lubricant goals in the 3 years since publication of the NRC Phase 4 review of U.S. DRIVE (NRC, 2013). In regard to fuel research accomplishments, the focus has clearly been on development of advanced fuel blends comprised of hydrocarbon and biofuel components. The research team led by Szybist at Oak Ridge National Laboratory has demonstrated a Reactivity Controlled Combustion Ignition (RCCI) engine using a gasoline/B20 dual-fuel strategy that meets the key ACECTT 2020 goal of 36 percent brake thermal efficiency (BTE) at 2,000 rpm and 20 percent peak load (Szybist et al., 2015). The RCCI operating range has been expanded to 75 percent of its theoretical maximum while maintaining low soot and NO_x emissions when using this fuel blend. This same team has also quantified the benefits of high-octane E30 (30 percent ethanol in gasoline) in downsized, turbocharged, direct-injection four-cylinder engines employing high compression ratios.

The team led by Zigler at the National Renewable Energy Laboratory (NREL; Zigler et al., 2015) has evaluated the derived cetane number of small fuel samples using an Ignition Quality Tester (IQT; ASTM Method D-6890). Over 388 pure compounds have been evaluated (122 by NREL using the IQT) and published in a compendium report (Yanowitz, 2014). The IQT has also been modified to provide kinetic data at gasoline combustion conditions of high pressures and temperatures. Octane reference fuels and gasoline/ethanol blends have been evaluated with the modified IQT for development of improved gasoline combustion kinetic mechanisms.

Sjoberg at Sandia National Laboratories has conducted a wide variety of analytical measurements in combustion engines operating on E0 to E30 gasoline/ethanol blends (Sjoberg, 2015). E30 blends are shown to be compatible with high-efficiency, boosted, direct-injection, stratified-charge engine performance. Sjoberg also concludes that ultralean SI operation requires end-gas autoignition for high-combustion efficiency.

A group at Lawrence Livermore National Laboratory (LLNL) led by Pitz has also developed a laboratory reactor that uses microliters of fuel to characterize the chemical kinetics of fuel combustion including gasoline, diesel, and diesel/biodiesel and ethanol/gasoline blends (Pitz et al., 2015). The "Micro-FIT" is a combustion tube that allows the evaluation of fuel ignition and flame extinction using only 5 to 20 μg/s of fuel. Multidimensional CFD simulations needed to understand fuel effects on lean/dilute direct injection, spark-ignition engines were conducted at Sandia National Laboratories (SNL) using data collected with the Micro-FIT at LLNL.

Mueller at SNL has conducted a detailed research program directed at reducing soot generation in diesel engines by modifying fuel composition and combustion strategies (Mueller, 2015). As part of a CRC program, a set of surrogate diesel fuels was developed, blended, and made available for laboratory test programs in combustion tubes and single-cylinder engines. An optical analytical technique, vertical-sheet, laser-induced incandescence, has been developed to analyze and quantify in-cylinder soot measurements. A diesel-biodiesel blend has been evaluated as an enabler for a leaner-lifted flame combustion strategy. Soot concentrations produced using this fuel/combustion combination were significantly reduced.

In regard to achievements in lubricant research, a DOE contract with Ashland Oil Corporation has led to the development of a heavy-duty engine oil formulation that provides greater than 2.0 percent fuel economy improvement relative to a CJ-4, SAE 15W-40 oil (Wu et al., 2015). SAE 5W-20 versions of this lubricant technology are predicted to provide greater than 2.0 percent fuel economy improvement in light-duty applications. Ashland has also developed an axle lubricant that can provide 0.7 percent improvement in fuel economy. These developments demonstrate significant progress in meeting the VTO goal of a 4.0 percent improvement in vehicle fuel economy provided by 2020 through reduction in driveline friction.

A research program led by Fenske at Argonne National Laboratory has focused on the development of methods for analyzing tribo-films that form on rubbing surfaces and affect friction and wear (Fenske, 2015). Techniques including focused ion beam electron spectroscopy and X-ray absorption near-edge structure techniques have been used to identify the degree of amorphous or crystalline nature of additive films on wear surfaces.

Significant Barriers and Issues That Need to Be Addressed

The DOE focus on petroleum fuels and lubricant research has developed into a well-structured portfolio of projects. There are no real barriers to conducting research on advanced fuel concepts under study at various national laboratories, suppliers, and OEM facilities. Instead there are only the challenges posed by DOE VTO management goals of identifying an advanced engine/fuel combination that reduces the nation's dependence on petroleum and increases the use of low-carbon, alternative fuel components and developing a plan for introducing such engine/fuel combinations in the commercial market. An additional challenge is to reduce driveline parasitic losses through the development of advanced lubricant formulations. The research programs currently under way are well structured to meet these challenges, but whether the program will meet its challenges according to the timeline proposed by the VTO is unknown.

In the view of the committee, meeting the timing of the goals identified by the DOE for advanced engine/fuel combinations, although well intended, will be difficult. The committee was told that the VTO plans to "downselect" a specific set of candidate fuel properties by 2017 and to demonstrate an optimized kinetically controlled engine/fuel system by 2025 (Howell, 2016). This is a very aggressive set of objectives. Although the current portfolio of projects will provide technical data to aid in making the selection, the process of choosing an optimized engine/fuel system will be difficult. Presumably, the Co-Optima program will play an important, if not defining, role in selecting such optimized engine/fuel systems. However, reaching consensus within a diverse group of government and commercial interests is going to be a severe challenge. Each potential engine/fuel combination will have technical, economic, and political benefits and drawbacks. It is not too early in planning to identify the process and criteria for selecting an optimum system. Finally, it is also not too early to formulate the plans for promoting the use of such an engine/fuel combination in commercial vehicles.

Response to Recommendations from Phase 4 Review

There were no recommendations dealing with petroleum-based fuels or lubricants in the NRC Phase 4 review.

Appropriate Federal Role

It is entirely appropriate for the federal government to be involved in research that affects two independent, major U.S industries, auto and energy. It is the committee's opinion that there is very little chance that a future advanced engine/fuel combination that meets national goals for efficiency and emissions will be universally accepted by all partners. The national laboratories can serve a critical role as arbiter of conflicting data and disagreements regarding potential benefits. The introduction of new fuels into the marketplace that must precede, or at least be in conjunction with, the introduction of advanced engines will occur only after the development and agreement on commercial fuel-quality standards. Participation of national laboratories personnel in these standards-setting processes would add significant technical resources and expertise to this effort and accelerate the standards-setting process. The Co-Optima initiative may help in reaching consensus on these challenging technical and economic issues, although more information is needed on how this program will deal with such issues.

Findings and Recommendations

Finding 3-7. The portfolio of projects assembled by the U.S. DRIVE Fuels Working Group (FWG) and focused on the development of advanced petroleum-based fuels is well structured. Engine tests are being conducted or planned to evaluate the efficiency and emissions of petroleum/biofuel blends in dilute/lean spark-ignited (SI) engines, clean diesel engines, and low-temperature combustion engine designs.

Finding 3-8. Laboratory and dynamometer tests conducted at national laboratories are contributing to progress in meeting specific U.S. DRIVE ACECTT goals. They are generating important data in support of development of combustion models using advanced fuels, different injection strategies, and combustion chamber designs.

Finding 3-9. DOE has set an aggressive timeline for identifying an "optimized kinetically controlled" engine/fuel system. The Co-Optima program will presumably help in developing the data to establish such an "optimized" system, but the DOE has not yet addressed how such a system would be implemented in the light-duty vehicle fleet.

Finding 3-10. DOE needs to provide greater explanation of how the Co-Optima program will be managed. How are the research projects at the national laboratories set to meet the technical needs of the U.S DRIVE OEMs under the Co-Optima program?

Recommendation 3-2. Following on presentations made to the committee, the Department of Energy (DOE) should further explain how the Co-Optima program will lead to the introduction of an optimum engine/fuel system in commercial practice. The introduction of high-efficiency, low-greenhouse gas internal combustion energy technology into the marketplace may require fuel formulations that are different from today's commercial fuels. Engine manufacturers will not introduce vehicles that utilize advanced combustion systems without the assurance that suitable fuels are available for the new combustion technology. Reaching consensus between the DOE's Co-Optima program and U.S. DRIVE on the concept of an optimum engine and fuel is necessary, but not sufficient. A plan for introduction of advanced combustion systems and fuels designed to increase transportation energy efficiency and reduce CO_2 emissions is required.

Biofuels

Background

As discussed in NRC (2015b) and other places, Congress established the Renewable Fuel Standard (RFS) in 2007, which set a goal of using 36 billion gallons of biofuels per year in transportation applications by 2022. To meet this goal, Congress has provided tax credits and incentives for biofuels production. These credits and incentives generally remain in effect.

Despite the federal government's desire for increased biofuel use, the only significant source of biofuels in use today is corn-derived ethanol. This ethanol is added to gasoline in the United States, mostly at a concentration of 10 percent. In early 2011, the EPA expanded a waiver to allow up to 15 percent ethanol in gasoline used to fuel 2001 and later model year light-duty vehicles. The EPA cannot force fuel stations to provide gasoline blends containing 15 percent ethanol without the approval of Congress, which at this time, it does not have. The use of 15 percent ethanol has been opposed by some global OEMs due to concerns over fuel system durability in older engines designed for 10 percent ethanol in gasoline. To date very little gasoline containing 15 percent ethanol has become available at commercial fuel pumps.

The commercial production of cellulose-derived ethanol, envisioned in the RFS, is slowly being realized.[4] Three different companies, Abengoa Bioenergy, POET, and DuPont have been producing ethanol from cellulosic feedstocks since either 2014 or 2015 (Abengoa Bioenergy, 2014; POET, 2014; Lane, 2014). The combined production of ethanol from these three plants could reach 60 million barrels per year. The plants are using technology that was developed in part at the NREL.

The use of bio-based butanol is also being considered for transportation use. Butanol has better gasoline blending and vapor pressure characteristics than ethanol, while still providing a significant octane boost. Butanol can be produced from renewable biomass, and it can be produced from refinery operations that produce excess C4 hydrocarbons. However, the production of butanol from petroleum light ends in the refinery would not be considered a renewable fuel.

The production of biodiesel (essentially fatty acid methyl ester [FAME] and other esters) is still minimal but continues to increase (National Biodiesel Board, 2016). Although the process for production of FAME is well defined, there is

[4] Cellulosic feedstock is primarily sourced from the structural components of plants and trees (e.g., trunks, branches, stems, stalks, leaves, and roots).

significant research and development being conducted on the identification and production of other oxygenated diesel fuel components that could provide benefits in clean diesel combustion applications.

Renewable "drop-in" fuels (fuels that are derived from biomass feedstock but processed into commercial fuel components using refinery operations) are the subject of research in both national and industrial laboratories (Peckham, 2014). National laboratory research programs are defining the performance of both wood-derived bio-gasoline and bio-reformates in engine tests. Wood-derived gasoline is produced by gasifying wood residue, compressing the resulting syngas, and then catalytically reforming the components of the gas stream (Farenback-Brateman et al., 2016). Bio-reformate components are produced from starch, sugar, or cellulosic feedstocks that are subjected to catalytic reforming with extra added hydrogen. Although the technology for producing "drop-in" biofuels is known, further development is needed to reduce production costs per unit of energy produced (Brown and Capareda, 2015).

Status of U.S. DRIVE Biofuel Goals and Targets

The U.S. DRIVE biofuel goals and targets are essentially the same as those described earlier in this chapter for petroleum-based fuels. If there is any difference, it is in the emphasis on the validation of biofuel components that might be blended into gasoline or diesel fuel to promote improved efficiency and reduced emissions.

The projects listed in Table 3-3 describe research that focuses not only on petroleum fuel chemical properties and combustion characteristics but also on biofuel characteristics and petroleum/biofuel blends that provide specific properties to aid different advanced combustion strategies. The projects listed are intended to address issues that will facilitate meeting ACECTT and FWG objectives and goals. In addition, as with petroleum projects, the U.S. DRIVE groups work closely with oil company representatives, OEM members, and national laboratory personnel in Coordinating Research Council research groups and committees.

Progress and Key Achievements

As indicated previously, most of the key achievements in regard to VTO fuel development projects since the last U.S. DRIVE review have involved the use of petroleum fuel/biofuel blends. Many of these projects were listed in Table 3-2 as part of the ACECTT's combustion, emission, and aftertreatment activities discussed earlier in the advanced engine combustion portion of this chapter.

One research area not covered is the work of McCormick and his team at NREL. This research project is investigating the possibility of using long-chain oxygenates derived from biomass as drop-in fuel components for blending with either gasoline or diesel fuel (McCormick et al., 2015). The oxygenate candidates are being solicited from suppliers and are being evaluated in laboratory tests to determine fuel blend properties. In addition, the fuels are also being evaluated in both diesel and gasoline direct-injection (GDI) engine tests. Twenty-four different oxygenates have been considered. Partial results have demonstrated that phenolics and esters improve fuel lubricity. Ethers and ketones have no effect on lubricity. There is no effect of oxygenates on oxidation stability except for phenolics that act as antioxidants. Early engine test results have shown no effect of oxygen on particulate matter emissions. GDI tests have shown that particulate matter and particulate numbers for gasoline/oxygenate blends follow the same trends as for hydrocarbon fuels.

Significant Barriers and Issues That Need to Be Addressed

Cost challenges represent a substantial barrier to use of biofuels in commercial fuel blends. In the case of drop-in fuels derived from biomass, it has been reported that the cost of production is a major deterrent confronting commercialization of these fuel components (Brown and Capareda, 2015). In the NRC Phase 4 U.S. DRIVE review, it was stated that the DOE target for achieving cost-effective drop-in fuels was 2017 (NRC, 2013). There have been no data presented that would indicate that this objective will be met on time. It appears that this is a challenge that will continue into the future.

One issue that does not seem to be sufficiently addressed in many of the VTO research projects is the issue of engine and fuel system long-term hardware durability with any new commercial fuel or fuel blend. Obviously, first meeting goals related to fuel efficiency and emissions is paramount. Furthermore, it is the responsibility of each OEM to decide if their products will survive for acceptable periods when operating on a new fuel. However, early tests within participating research and development laboratories could provide indications of the significance of any durability issues. Early tests of oxygenated fuel effects on fuel lubricity are a good example of this effort (McCormick et al., 2015). EPA providing a waiver for 15 percent ethanol use in commercial gasolines is an example of the resistance that can be generated by both OEM and petroleum distributors when a new fuel blend is authorized without sufficient early durability testing.

Appropriate Federal Role

The programs funded by the VTO that include research on biofuels in support of U.S. DRIVE objectives are well organized. If the United States is to achieve its stated goal of using 36 billion gallons of biofuels in transportation applications, alternatives other than corn-based ethanol in gasoline need to be developed and commercialized. Since the federal government is the author of the Renewable Fuel Standard designed to reduce the use of petroleum-based fuels and greenhouse gases from the transportation sector, the government has an appropriate role in conducting research and development in support of the achievement of its stated goals.

Findings and Recommendations

Finding 3-11. The use of ethanol as a blending component in petroleum-based fuels is being well researched and evaluated for use in advanced combustion concepts. There is less emphasis on the benefits of other low molecular weight alcohols (higher molecular weight than ethanol, but lower than C7 alcohols, e.g., butanol, pentanol, etc.). Although these alcohols would most likely be produced from petroleum feedstocks, they could provide benefits in fuel blending and advanced combustion strategies.

Finding 3-12. "Drop-in" fuels derived from biomass can be produced from recognized unit operations within a refinery, but cost is currently a challenge for use of these components in commercial fuel blends.

Finding 3-13. The effect of biofuel components on engine and fuel system durability is as important as their effect on engine efficiency and emissions.

Recommendation 3-3. It is recommended that in support of ongoing U.S. DRIVE, Coordinating Research Council, and Co-Optima fuel evaluation programs designed to improve internal combustion engine efficiency and reduce greenhouse gas emissions, other offices within the Department of Energy such as the Bioenergy Technologies Office continue to provide support for development of cost-effective manufacturing processes for converting biomass to liquid "drop-in" hydrocarbons compatible with gasoline and diesel fuel blends. Achieving lower cost manufacturing goals would provide U.S. DRIVE with a new pathway to meeting their objective of increasing the amount of biomass-derived fuels used in commercial fuel blends.

Recommendation 3-4. The Department of Energy's national laboratories should systematically evaluate the effect of all potential biofuel components on engine and fuel system durability as well as their effect on engine efficiency and emissions.

Natural Gas

Background

A detailed review of the use of natural gas in medium- and heavy-duty trucks has recently been published by the NRC (2014). The reader is referred to this review for information regarding the justifications, availability, issues, and benefits related to the use of natural gas as a transportation fuel. As described earlier in this chapter, recent developments in "fracking" and drilling technologies have greatly increased the supply and future estimates of reserve gas available in the United States (EIA, 2016a). The current production of natural gas using these advanced extraction techniques is so great that the Energy Information Administration has predicted that the United States will become a net exporter of gas by 2018 (EIA, 2016a). In addition, current natural gas supplies have driven the price of natural gas to its lowest level in recent history. This has resulted in much greater use of natural gas in electricity generation and slightly greater use as a medium- and heavy-duty truck fuel. In comparison, the use of natural gas in U.S. light-duty vehicle applications remains very small.

OEMs, both domestic and foreign, have been producing and marketing natural gas vehicles that meet fuel economy and emissions regulations for over three decades. Despite the requirement for large, expensive fuel tanks needed to provide acceptable driving "range" and that also create onboard packaging and storage challenges in vehicles, natural gas vehicles have achieved significant penetration in vehicle populations in numerous countries that have a suitable natural gas vehicle fueling infrastructure (e.g., Italy, Argentina, etc.). The reasons for limited sales of such vehicles in the United States can be attributed to three economic rather than technical issues:

1. New vehicle costs being above gasoline-fueled vehicles because of more expensive emissions control components and large bulky storage tanks;
2. Limited, high-cost natural gas refueling infrastructure for private citizens; and
3. Poor resale value for fleet owners that use centralized refueling facilities when they try to sell used natural gas vehicles to private citizens who have no access to refueling facilities.

With the exception of further materials research to lower the costs of emissions control components and storage tanks, there is little that the VTO can support in research programs that will solve these problems. For this reason, it is not surprising that there are no projects listed in Table 3-3 that focus on the use of natural gas as a light-duty vehicle fuel.

There is the possibility of using natural gas in a Fisher-Tropsch reaction to produce drop-in hydrocarbon fuels (either gasoline or diesel fuel). Currently there are large Fisher-Tropsch plants that are producing commercial products, but most of these are relatively high-cost lubricating base oils or chemical feedstocks. There are also mini Fisher-Tropsch plants that are planned for use at stranded gas deposits (not connected to a pipeline). Conversion of "gas to liquids" at these stranded sites allows liquid feedstocks to be transported by truck as opposed to by pipeline (Peckham, 2014). However, such mini Fischer-Tropsch plants are also challenged by unfavorable economics.

Status of U.S. DRIVE Natural Gas Goals and Targets

There are no known VTO initiatives in support of the U.S. DRIVE Partnership that utilize natural gas in their research projects. There are as many as 13 projects investigating natural gas as a transportation fuel within the DOE Advanced Research Projects Agency-Energy (ARPA-E) Methane Opportunities for Vehicular Energy initiative, but it does not appear that these projects are driven by U.S. DRIVE goals and objectives at this time (ARPA-E, 2012).

Significant Barriers and Issues That Need to Be Addressed

The largest barrier to the use of natural gas as a transportation fuel for light-duty vehicles is the lack of a refueling infrastructure for private vehicle owners. A population of refueling stations that would meet individuals' needs for both local and long-distance trips could remove a good portion of the resistance to purchase light-duty vehicles fueled by natural gas, as it has in many countries around the world. However, it is worth considering whether the development of a hydrogen infrastructure and public refueling capability for use with fuel cell vehicles will create economic resistance to development of a similar natural gas refueling network for internal combustion engine-equipped vehicles. It is reasonable to ask if the United States will support the development of two public gaseous fuel infrastructure networks.

Response to Recommendations from Phase 4 Review

> ***NRC Phase 4 Recommendation 3-2.*** U.S. DRIVE should make an assessment of whether natural gas can be an enabler for achieving the advanced combustion modes currently being pursued in its research portfolio.
>
> ***Partnership Response.*** Over the last several years, the Partnership has considered but elected not to include natural gas in its technical scope, given its focus on precompetitive research. However, with an emphasis on advanced combustion engine technologies, the Advanced Combustion and Emission Control (ACEC) Technical Team includes natural gas in the latest version of its roadmap. Specific roadblocks to the widespread adoption of natural gas as a vehicle fuel include ignition energy requirements and low-temperature methane oxidation catalysts, as well as onboard storage and infrastructure. In terms of an enabler for advanced combustion modes, the ACEC roadmap identifies natural gas as a fuel for the dilute combustion mode and down-sized boosted engines, which may spur more research into its use for other advanced combustion modes.

Committee Assessment of Response to 3-2. U.S. DRIVE also supplied the committee with extensive documentation that supports the summary statement given in their response. The committee thanks the U.S. DRIVE Partnership for the comprehensiveness of their response and is pleased that they have evaluated compressed natural gas in terms of its technical and economic viability for use in the light-duty mobility market.

Given that there are ample supplies of natural gas available in the United States at this time, coal is slowly being displaced by natural gas in electrical power generating stations because of its relatively low cost and its value as a method for reducing GHG emissions (EIA, 2016a) during electricity production. Because of the infrastructure needs that would be required for widespread use of natural gas in transportation applications, it is reasonable to think of natural gas, if it can be produced and transported without stray methane emissions, as the most clean and efficient fuel for use in electricity generation.

Appropriate Federal Role

It is entirely appropriate for the federal government to support research on methods for improving efficiency and reducing emissions related to the use of natural gas, if it believes that it is a viable option as at transportation fuel. There are valid applications for use of natural gas as a transportation fuel for medium- and heavy-duty trucks (Class 6 through Class 8, including metro buses, school buses, garbage trucks, and over-the-road tractor trailers) with access to dedicated refueling facilities.

Findings

Finding 3-14. There are no research projects at this time funded by the VTO in support of U.S. DRIVE goals and objectives that are focused on the use of natural gas as a fuel for advanced combustion systems. There are projects being conducted within ARPA-E focused on use of natural gas as a transportation fuel, although it does not appear that this focus is in response to U.S. DRIVE goals and objectives.

Finding 3-15. Liquid "drop-in" fuels could be produced from Fisher-Tropsch processes that use natural gas as a feedstock.

FUEL CELLS

Introduction and Background

Hydrogen fuel cell vehicles (HFCVs) have been in a development phase by the major automotive companies for decades. Their attractiveness when compared to current internal combustion engine technology is based on the direct conversion of chemical (hydrogen) to electrical energy via an electrochemical process, thereby resulting in reduced environmental impact provided that the hydrogen is derived from "green" primary energy sources. The efforts to develop HFCVs by the major automotive companies have been significant as is evident from the magnitude of the investments made by the individual automotive OEMs, the number of patents issued (DOE, 2016e), and the engineering accomplishments to date (U.S.

DRIVE, 2015). Notably, within this review period, a number of foreign OEMs (Toyota, Hyundai, and Honda) either have initiated membrane-based fuel cell vehicle sales or leases to the general public in the United States or have announced that vehicles will be available within the next few years. General Motors, Ford, and Fiat Chrysler, all three U.S. DRIVE Partnership members, do not currently have vehicle offerings, yet GM has been cited in the open literature as stating they are "on track" to produce their Gen 2 HFCV by 2020.[5] Recent activities by the aforementioned OEMs, foreign and domestic, demonstrate that HFCVs are in the late stages of development and are, or are soon to be, ready for customer engagement, albeit at a modest level owing to limited production volume and hydrogen delivery and refueling infrastructure issues. The U.S. DRIVE Partnership, which is focused on coordination and communication of activities funded by the FCTO within EERE at the DOE and the OEMs, has been and remains fully engaged in all aspects of HFCV development.

A key interest in electric vehicles (battery or fuel cell powered) has been the potential of this technology to ultimately achieve near-zero carbon emissions (Satyapal, 2016) if renewable energy sources such as wind, solar, hydroelectric, or other non-fossil-fuel-based sources are used to generate the hydrogen needed to fuel the HFCVs. If achieved, this would facilitate meeting the low-emission mandates created over the last two decades. Even if fueled by hydrogen derived from natural gas, life-cycle emissions of carbon dioxide and pollutants would still be lower than those from current internal combustion engine emissions (DOE, 2015b).

Despite foreign OEM HFCVs entering an introductory customer-engagement phase, and with General Motors expected to follow in the next few years, challenges specific to fuel cell technology remain (Figure 3-3). Such challenges have been outlined in prior NRC program reviews (Phases 1 to 4) and many have been resolved. Meeting cost and fuel cell durability targets simultaneously remains the most critical barrier to overcome if HFCVs are to become viable, both technically and commercially.

Notable Changes since the NRC Phase 4 Report

In addition to the initial availability of HFCVs, another notable change since the NRC Phase 4 report is that the DOE has recently created a Fuel Cell Consortium for Performance and Durability (FC-PAD) to coordinate activities conducted at national laboratories, in conjunction with industry and universities, in the areas of electrocatalysts and supports, electrode layer, ionomers, modeling and validation, operando (i.e., operating conditions) evaluation, and component characterization. The overall goals of the consortium are to advance the performance and durability of proton exchange membrane (PEM) fuel cells at a precompetitive

[5] See, e.g., http://www.usatoday.com/story/money/cars/2017/01/30/general-motors-honda-fuel-cell-deal/97240096/ or http://www.autonews.com/article/20161011/BLOG06/310119999/fuel-cell-puzzle-comes-together.

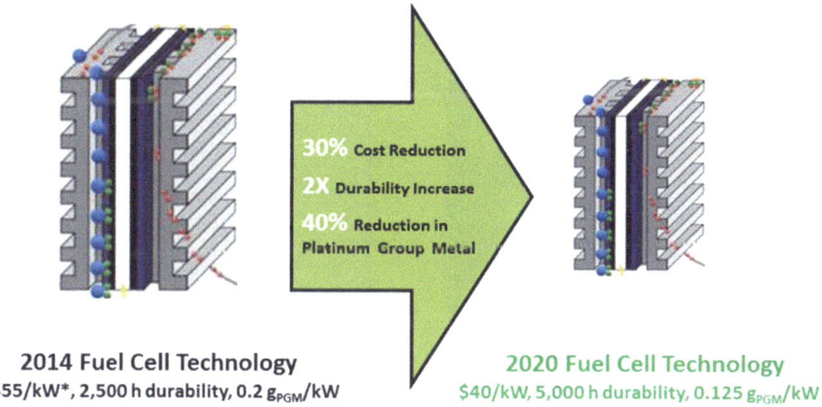

FIGURE 3-3 Fuel cell challenges. SOURCE: Satyapal (2016).

level in order to facilitate their commercialization, develop the knowledge base, and optimize structures for more durable, high-performance PEM fuel cell components. Reducing costs and improving high current density performance at low platinum (Pt) loadings are also anticipated outcomes. The committee supports the creation of FC-PAD and its focus on key electrochemical electrode issues.

The newly formed DOE-facilitated ElectroCAT consortium is in its formative stages and like the FC-PAD, it is to be national laboratory-centric with projects involving industry and academia that focus research activities on non-platinum-group-metal fuel cell catalysts.

Budgets

Because the Partnership is not a funding entity, a direct link does not exist between the identified needs developed by the fuel cell technical team (FCTT) and the funds allocated for specific R&D by DOE. As a result, the entire FCTO fuel cell budget (Figures 3-4 and 3-5) should be viewed as contributing to technology development for the FCTT targets (see Table 3-6, later) set forth by the members of the Partnership. The current DOE budget for fuel cell development of approximately $35 million per year represents about 30 percent of the overall EERE FCTO appropriation and has remained stable in recent years. Additional funding of related technology takes place in other DOE offices such as Basic Energy Sciences, ARPA-E, and the Small Business Innovation Research (SBIR) program. Appropriately, as shown in Figure 3-5, catalysts and electrodes receive the largest allocation, yet funding of membrane topic areas has been decreasing in recent years. The most significant increase appears to be in the "Fuel Cell Operation & Performance" category followed by a slight increase in "Testing and

Key Activity	FY 15	FY 16	FY16
	($ in thousands)		
	Approp.	Request	Approp.
Fuel Cell R&D	33,000	36,000	35,000
Hydrogen Fuel R&D[1]	35,200	41,200	41,050
Manufacturing R&D	3,000	4,000	3,000
Systems Analysis	3,000	3,000	3,000
Technology Validation	11,000	7,000	7,000
Safety, Codes and Standards	7,000	7,000	7,000
Market Transformation	3,000	3,000	3,000
NREL Site-wide Facilities Support	1,800	1,800	1,900
Total	97,000	103,000	100,950

Office	FY 2015
EERE	$97.0M
Basic Science	$18.5M
Fossil Energy, SOFC	$30.0M

FY 2015 DOE Total: ~$150M

Number of Recipients funded from 2008-2015	
Industry	>110
Universities	>100
Laboratories	12

FIGURE 3-4 Department of Energy's Fuel Cell Technologies Office budget. NOTE: SOFC = solid oxide fuel cell. SOURCE: Satyapal (2016).

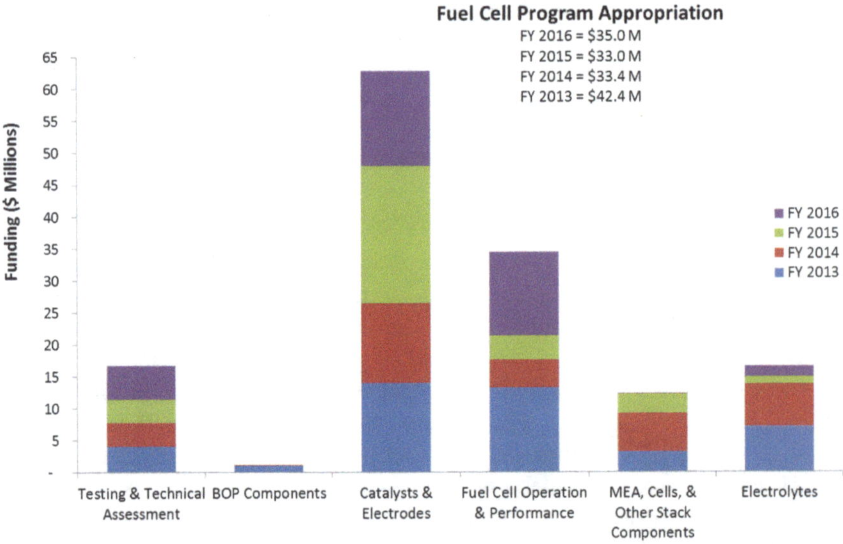

FIGURE 3-5 Department of Energy's Fuel Cell Technologies Office topic appropriations. SOURCE: Masten and Papageorgopoulos (2016).

Technical Assessment." It can be concluded that cost and durability issues are the primary focus areas. With that said, this conclusion draws attention to the delineation between near- and long-term projects incorporating the premise that the Partnership focuses only on identifying and addressing precompetitive concepts.

The ongoing DOE-funded projects that could impact Partnership-derived targets are presented in Table 3-5.

It is important to point out that in addition to the DOE funding, the OEMs are using their own internal financial resources to advance their respective proprietary technologies and specific needs. The DOE-supported precompetitive projects listed in Table 3-5 being conducted by national laboratories, academia, and private industry allow for critical resources at the OEMs to focus on getting a product to market.

Current Status vis-à-vis Goals and Targets

Three of the key 2020 technical targets set forth by the Partnership have been met as reported by DOE—power density, start temperature, and specific power (Table 3-6 and Figure 3-6). Though progress has been made in meeting the remaining targets, two continue to remain difficult to meet simultaneously, those being cost and durability (Wimmer and Gazelle, 2016). These were also identified in the NRC Phase 4 review report in 2013 (NRC, 2013). Since then, progress toward meeting the cost and durability goals has been minimal (Satyapal, 2016). It should be noted that the actual numbers achieved by the OEMs are generally not publicly available and therefore it is difficult to know or even monitor actual advancements relative to the FCTT targets. Furthermore, lifetime and durability testing requires significant blocks of time to perform; therefore, any currently reported data by the DOE may not be on the most recent generation of a given OEM's HFCV. The most recent non-OEM reported data by NREL using on-road HFCVs, including one U.S. OEM (GM), generated from tests conducted between 2012 and 2015 (Kurtz et al., 2015), does cite an increase in lifetime, but still well shy of the 5,000-hour target. An on-site visit to GM's fuel cell development operations in Pontiac, Michigan, by the fuel cell subgroup of the committee confirmed that significant advancements have been made toward the cost and durability targets.[6] GM disclosed that they had achieved stack lifetimes (durability) exceeding DOE targets in laboratory tests. With respect to cost reduction, GM confirmed a press report (Truett, August 2015) of platinum group metal (PGM) stack loadings of 10 grams, reduced from 20 to 30 grams. It remains to be demonstrated that the durability reported to the committee can be achieved with the lower catalyst loadings.

Significant progress has been made by the national laboratories in gaining a better understanding of the fundamental chemical and materials science and engineering, which can impact cost (platinum loading) and durability.

[6] The committee gathered this information during a visit to General Motors on June 21, 2016.

TABLE 3-5 List of Selected Department of Energy-Funded Fuel Cell-Related Projects

Presentation Title	Organization
Extended, Continuous Pt Nanostructures in Thick, Dispersed Electrodes	National Renewable Energy Laboratory
Nanosegregated Cathode Catalysts with Ultra-Low Pt Loading	Argonne National Laboratory
Contiguous Pt Monolayer Oxygen Reduction Electrocatalysts on High-Stability, Low-Cost Supports	Brookhaven National Laboratory
Fuel Cells Systems Analysis	Argonne National Laboratory
Fuel Cell Vehicle and Bus Cost Analysis	Strategic Analysis, Inc.
Characterization of Fuel Cell Materials	Oak Ridge National Laboratory
Neutron Imaging Study of the Water Transport in Operating Fuel Cells	National Institute of Standards and Technology
Fuel Cell Fundamentals at Low and Subzero Temperatures	Lawrence Berkeley National Laboratory
Effect of System Contaminants on Polymer Electrolyte Membrane Fuel Cell Performance and Durability	National Renewable Energy Laboratory
Technical Assistance to Developers	Los Alamos National Laboratory
The Effect of Airborne Contaminants on Fuel Cell Performance and Durability	Hawaii Natural Energy Institute
Fuel Cell Technology Status: Degradation	National Renewable Energy Laboratory
Roots Air Management System with Integrated Expander	Eaton Corp.
High-Performance, Durable, Low-Cost Membrane Electrode Assemblies for Transportation Applications	3M
Rationally Designed Catalyst Layers for Polymer Electrolyte Membrane Fuel Cell Performance Optimization	Argonne National Laboratory
Non-Precious-Metal Fuel Cell Cathodes: Catalyst Development and Electrode Structure Design	Los Alamos National Laboratory
Advanced Ionomers and Membrane Electrode Assemblies for Alkaline Membrane Fuel Cells	National Renewable Energy Laboratory
New Fuel Cell Membranes with Improved Durability and Performance	3M
Advanced Hybrid Membranes for Next-Generation Polymer Electrolyte Membrane Fuel Cell Automotive Applications	Colorado School of Mines

TABLE 3-5 Continued

Presentation Title	Organization
High-Throughput Synthesis, Oxygen Reduction Reaction Activity Modeling, and Testing of Non-Platinum-Group-Metal Polymer Electrolyte Membrane Fuel Cell Cathode Catalysts	Argonne National Laboratory
Novel Non-Platinum-Group-Metal Catalysts from Rationally Designed Three-Dimensional Precursors	Argonne National Laboratory
Platinum-Group-Metal-Free Catalysts for Polymer Electrolyte Membrane Fuel Cells	Savannah River National Laboratory
High-Performance and Durable Low-Platinum-Group-Metal Cathode Catalysts	Pacific Northwest National Laboratory
Magnetic Annealing of Pt-Alloy Nanostructured Thin-Film Catalysts for Enhanced Activity	Oak Ridge National Laboratory
High-Conductivity, Durable, Anion-Conducting Membranes	Oak Ridge National Laboratory
Advanced Hydroxide-Conducting Membranes	Los Alamos National Laboratory
High-Temperature and Low-Humidity Membranes	Sandia National Laboratories
Engineered Low-Pt Catalyst Layers	Los Alamos National Laboratory
Semi-Automated Membrane Electrode Assembly Fabrication with Ultra-Low Total Platinum-Group-Metal Loadings	Brookhaven National Laboratory
Durability Improvements through Degradation Mechanism Studies	Los Alamos National Laboratory

SOURCE: Cooper (2016a).

Cost

As in previous reviews of this Partnership, the cost target of $40/kW has proven to be difficult to reach, the current value being $53/kW (see Figure 3-7; Satyapal, 2016). Little progress has been made since 2010, further emphasizing that research must continue so as to offer new technical approaches that result in lower cost options. Such cost reductions must come from simplification of the system architecture as well as from new component solutions and operational approaches. System and stack costs are presented in Figure 3-8 as a function of volume; though interesting, it is not clear to the committee as to why low volume cost data (<500k units) are of any value.

As the state of maturity of fuel cell technology is still early, cost issues can be addressed by new technical solutions and approaches and less so by volume manufacturing. It is also important to emphasize that technical initiatives related to cost reduction and durability are inter-related, leading to long-term, complex,

TABLE 3-6 Fuel Cell System and Component Status and Targets

Progress 2012-2016				
Characteristic	Units	2012 Status	2016 Status	2020 Target
Cost	$\$/kW_{net}$	55	53	40
Power density	W/L	400	640	650
Specific power	W/kg	400	659	650
Durability	h	2,500	3,900	5,000
MEA performance at 0.8 V	mA/cm^2	160	240	300
PGM total content	g/kW	0.19	0.16	0.125
Catalyst mass activity	A/mg_{PGM}	0.24	0.5	0.44

SOURCE: Masten and Papageorgopoulos (2016).

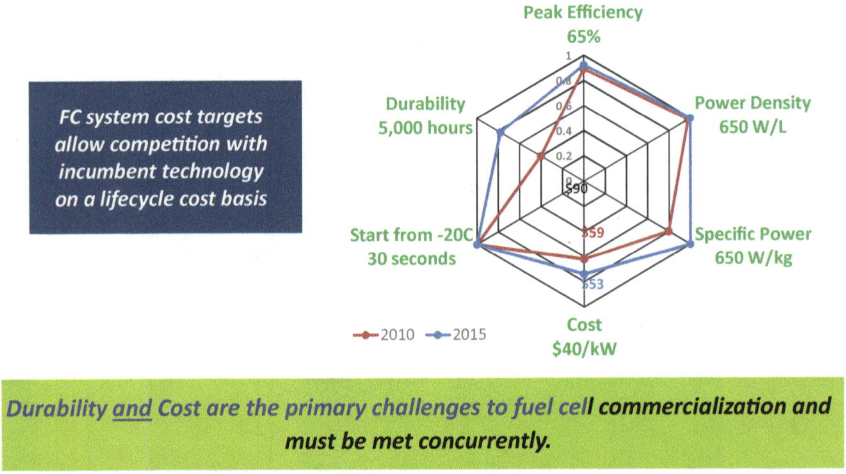

FIGURE 3-6 Department of Energy spider chart of fuel cell targets. SOURCE: Masten and Papageorgopoulos (2016).

multivariable R&D projects. Many such efforts have been in place for years at the national laboratories, in academia, and private industry funded by the DOE. It is imperative that such efforts continue, as it is anticipated that ultimately, significant advancements will emerge from such work.

A challenging issue for the committee is that the cost estimates made by third parties to evaluate progress toward meeting the cost target may not accurately describe the actual OEM costs. To further add uncertainty to the cost estimates, the OEMs may have already developed their own proprietary solutions either

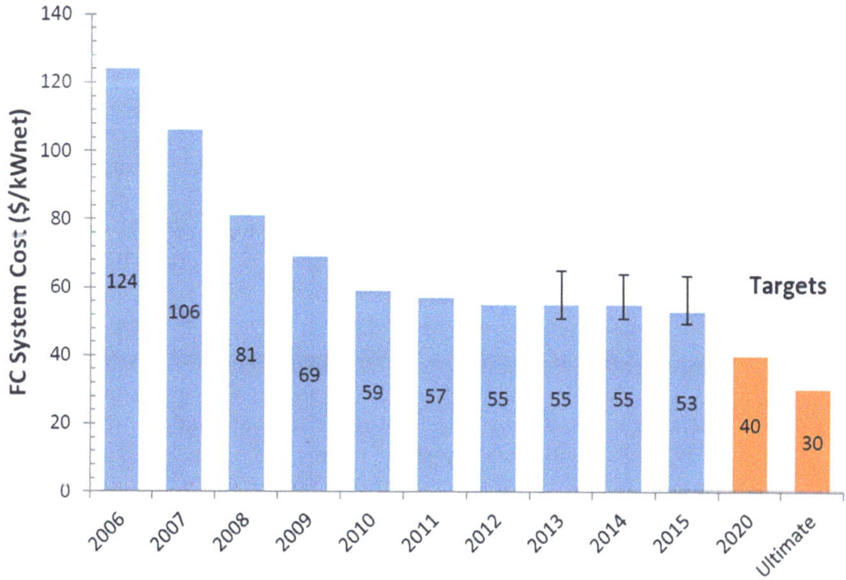

FIGURE 3-7 Estimates of fuel cell system cost at a production volume of 500,000 units per year. SOURCE: Satyapal (2016).

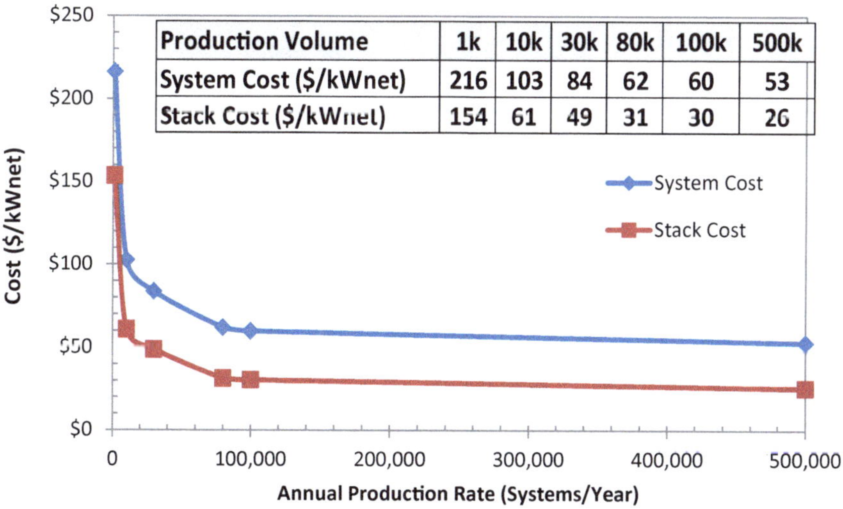

Production Volume	1k	10k	30k	80k	100k	500k
System Cost ($/kWnet)	216	103	84	62	60	53
Stack Cost ($/kWnet)	154	61	49	31	30	26

FIGURE 3-8 Fuel cell cost as a function of production volume. SOURCE: Masten and Papageorgopoulos (2016).

internally or together with the supply chain as DOE-funded projects are not likely to move at the same pace as the OEM internal development efforts. In a recent Strategic Analysis cost assessment (James and Spisak, 2012), selected components are used in the analysis, some of which are still being reported as development efforts by the private sector and are unproven. This adds a degree of risk to the credibility of cost assessments. As in the prior review, direct communication with one U.S. OEM indicated that even with this uncertainty, they are comfortable with the ultimate target of $40/kW. With that said, there were concerns regarding the differences between the OEM and the DOE-funded cost analyses based on the technology, systems, and operational approaches.

The sensitivity analyses reported by the DOE indicate that the catalyst loading, stack plates, and the air compressor are the components that have the largest impact on cost (Figure 3-9). Unique to fuel cells is that the current density and stack voltage can be controlled so as to impact performance and overall efficiencies. This characteristic will dictate design features that include cell area (plate area), the number of cells, as well as other components such as the radiator, air compressor, and blower sizing. As a result of the interdynamics of how the stack operates and behaves, the DOE has been appropriately funding studies and development activities in these areas, including system operating modes and systems architecture relationships, which to some extent depend on catalyst loadings and membrane and electrode characteristics. As presented in Figure 3-9, the number

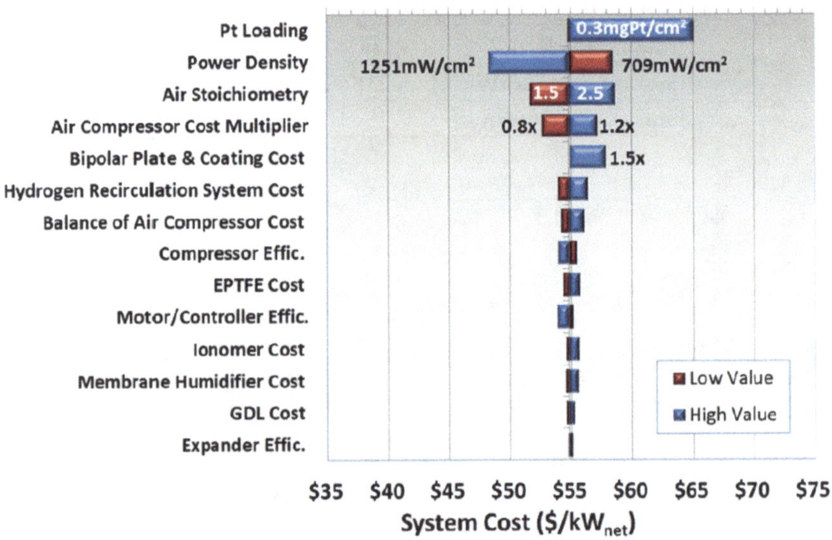

FIGURE 3-9 Sensitivity analysis of components that affect fuel cell cost. SOURCE: Masten and Papageorgopoulos (2016).

one issue is the platinum content used in the catalyst layers. Currently, as reported by the DOE, approximately 20-30 grams of platinum per stack are used, a quantity significantly higher than the target. The objective is to reduce the Pt content to the target loading set by the Partnership of 0.125 grams of Pt/kW. Lower catalyst loadings work but at the expense of durability, as previously discussed. In response to this ongoing challenge, the DOE continues to fund a number of academic, industrial, and national laboratory catalyst, cell and stack performance, architecture, and systems analysis projects. Newly generated solutions will address these complex interactions that will ultimately translate to cost reductions without sacrificing durability. GM suggested to the committee fuel cell subgroup that a DOE-funded study to generate an understanding of durability issues of low Pt levels could provide valuable insight. The committee agrees with this. The recently formed national laboratory-based consortium described above will attempt to determine the membrane and electrode degradation mechanisms, and develop new catalyst solutions. The ElectroCAT consortium will address lower cost catalyst and electrode layer technologies using non-PGM catalysts. Both of these activities are warranted and long overdue and will hopefully impact cost.

In summary, the committee feels that the cost assessments by third parties must be addressed with the OEMs directly in a way that does not compromise their respective proprietary and confidential technologies and business plans. Cost estimates must be based on an understanding of system architecture, mode of operation, and especially end-of-life specifications as such knowledge will impact the component life assessments and cost requirements. At the current state of vehicle development, it is unlikely the three U.S. OEMs would be willing to share such sensitive information. Thus, it is not clear as to how third-party estimates are of value. Last, the DOE-supported analyses of very low volume HFCV production is of little value.

Durability

Stack durability (target: 5,000 hours) as presented by DOE in Table 3-7 and Figure 3-10, like cost, has been met but only at the expense of higher catalyst loadings. Since the NRC Phase 4 review, average stack life reported to the committee by the DOE based on on-road NREL data (Kurtz et al., 2015) has increased to an average of 3,900 hours from six HFCV on-road tests. This figure is up 2,500 hours from that reported in the NRC Phase 4 review. With recent OEM HFCV advancements, including new technologies, new modes of operation, and from proprietary accelerated testing by the OEMs themselves, it is not clear if the various non-OEM organizations performing lifetime and stack performance tests are using the same hardware and test protocols, performance, and end-of-life metrics as the OEMs, and hence these numbers may not be representative of the state of the art.

Repeated themes throughout this review are that stack components are complex and that how they are assembled into stacks and then operated impact stack

TABLE 3-7 Department of Energy Reported Membrane Electrode Assembly (MEA) Performance and Durability Metrics

Technical Targets: Membrane Electrode Assemblies			
Characteristic	Units	2015 Status	2020 Targets
Cost	$\$/kW_{net}$	17	14
Durability with cycling	Hours	2,500	5,000
Start-up/shutdown durability	Cycles	—	5,000
Performance at 0.8 V	mA/cm^2	240	300
Performance at rated power (150 kPa abs)	mW/cm^2	810	1,000
Robustness (cold operation)	—	1.09	0.7
Robustness (hot operation)	—	0.87	0.7
Robustness (cold transient)	—	0.84	0.7

NOTE: 5,000 hours of operation under simulated vehicle power cycling and shutdown/start-up cycling with less than 10 percent loss in rated power. Specifically, developing MEAs with state-of-the-art catalysts that demonstrate performance greater than 1 W/cm^2 with Pt loading less than 0.125 mg/cm^2.
SOURCE: Allendorf and Borup (2016).

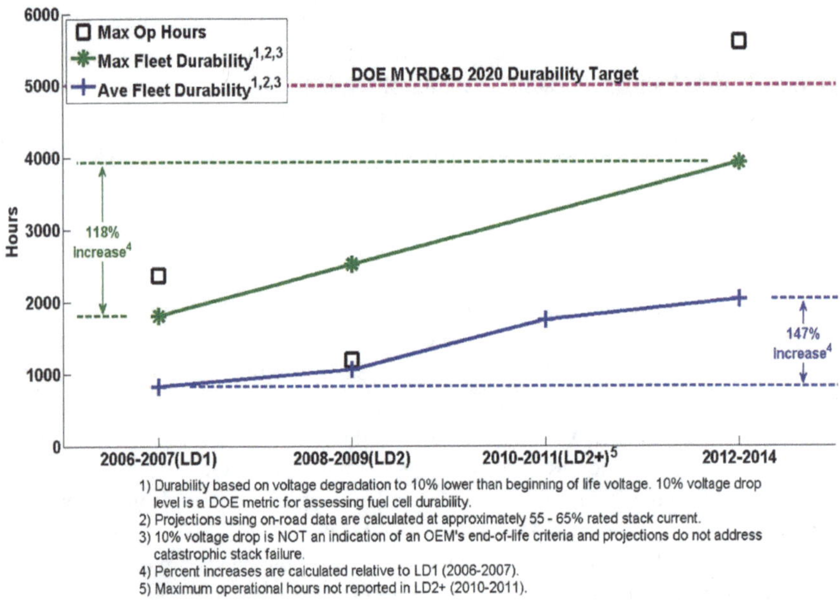

FIGURE 3-10 Fuel cell stack durability. SOURCE: Masten and Papageorgopoulos (2016).

durability. For example, membrane as well as catalyst ionomer hydration characteristics, electrode layer structure including porosity, composition, conductivity (electrical and ionic), reactant diffusion rates through the membrane, impurities, and many additional parameters can impact performance and durability. Operating modes of the stack can and will affect the chemical, mechanical, and physical nature of all of the above, thereby making control systems critical in developing a viable, long-life, durable stack. Each OEM is addressing these issues with their respective proprietary approaches. As reported to the fuel cell subgroup and not available to the public, GM, through extensive investments in infrastructure testing and analyses methods, has indicated significant advancements in stack durability. With that said, GM indicated that DOE support of new electrolytes for the membrane as well as the ionomer, and related topics, though long term, are warranted and should continue because membranes and catalysts are critical, and if possible, with increased levels of funding.

The most recent data in the public domain as presented to the committee are shown in Table 3-7 and Figure 3-10. Of particular interest is the durability with cycling data reported in hours in Table 3-7 and the average fuel cell operation data presented in Figure 3-10, also reported in hours. From these data, it is clear the durability of the membrane electrode assemblies (MEAs) of 2,500 hours is substantially above the average fleet durability lifetime reported (see Table 3-7 and Figure 3-10 references) to be approximately 2,000 hours. The fact that there is a difference in the two figures is anticipated, as actual vehicle data will include the effects of all aspects of the "system" on fuel cell stack performance, unlike data derived from single cell and stack testing in a laboratory. Although it is encouraging that the maximum fleet durability data are approaching 4,000 hours, there has been little change in the average fleet durability data since 2010 as presented in Figure 3-10. The information derived from the fundamental studies of MEAs will continue to provide valuable insight and guidance to the OEMs and the MEA suppliers as will the OEM vehicle testing from a systems and actual on-road perspective. It is anticipated that the recently formed FC-PAD consortium will contribute to the development a 5,000-hour-life MEA.

As previously mentioned, the FC-PAD consortium is chartered to address the durability question from a fundamental science and engineering perspective, with efforts from OEMs, industry, and academia. It will be up to DOE using the FC-PAD consortium to allocate funding in an effective way to make this a relevant program. At the minimum it will help coordinate the project portfolio.

Assessment of Progress and Key Achievements of the Partnership

Understanding fuel cell stack performance, the subsystem where the electrochemical processes take place, is complex, as it is a function of the membrane characteristics, catalysts, stack design, environment, and the overall fuel cell system architecture and mode of operation. Specifically, the durability, as stated previously, is predominantly a function of the chemical stability of the catalyst, catalyst layer composition, membrane content, membrane physical and chemical stability, and operating environment. Work on carbon-free supported catalysts, chemically and mechanically stable membranes that can operate at a lower humidity, and the impact of ionomer properties and interfacial catalyst layers and membrane stability, is essential and has continued to receive support in this review period as recommended in prior NRC reviews.

In consideration of the preceding, there are a number of promising projects funded by the DOE that could make a significant impact on commercialization efforts by member OEMs including those listed here.

The FCTT of the Partnership recognizes that catalysts represent roughly 50 percent of the membrane-based stack cost and that reducing catalyst cost remains a key challenge to lowering the cost of fuel cells. With this in mind, the FCTT has encouraged partners to focus on reducing the PGM content of fuel cell catalysts, increasing the activity of PGM catalysts without sacrificing their durability, and exploring non-PGM catalysts and novel catalyst supports. The efforts toward these objectives have resulted in notable progress, as exemplified by the following illustrations:

- Argonne National Laboratory (ANL) has developed a method for meso-structuring PtNi thin films so that they exhibit 20x the specific activity of Pt nanoparticles supported on carbon.
- GM has found that slow de-alloying Pt_3Ni nanoparticles to remove Ni and create a Pt skin results in a mass activity higher than 0.44 A/mg_{Pt} for cathode loadings of approximately 0.1 mg_{Pt}/cm^2.
- ANL and Lawrence Berkeley National Laboratory (LBNL) have developed methods for producing Pt_3Ni nanoframes that when supported on carbon exhibit an oxygen reduction reaction activity of over 5 A/mg_{Pt} at 0.9 V in 0.1 M $HClO_4$ and show little loss in activity after 10^4 cycles.
- Brookhaven National Laboratory (BNL) has investigated the performance of catalysts comprised of a Pt monolayer deposited over a PdNi core that exhibits good activity when incorporated into a membrane and electrode assembly.
- ANL has used structure-property modeling methods and new synthesis routes to produce new compositions of non-PGM catalysts.
- Los Alamos National Laboratory (LANL) has developed an engineered ionomer topology to facilitate proton transport, which when coated over Pt supported on multiwalled carbon nanotubes as a thin film led to enhanced oxygen transport to the Pt nanoparticles and targeted polarization curves at a Pt loading of 0.05 mg_{Pt}/cm^2.

Progress has also been made in developing a better understanding of the electrode structure, composition, performance, and degradation mechanisms as exemplified by the following two citations. Myers (2016) continues to build on the knowledge regarding catalyst support degradation mechanisms as well as the behavior of platinum under all operating conditions. More (2016) has focused on the ionomer used in the electrode layer. Following the recommendations of the NRC Phase 4 report, activities in this area have been added to include projects addressing non-PGM catalysts, enhanced imaging methods of the ionomers in the electrode and gas diffusion layers, ultra-low catalyst loadings, especially for the cathode, continuation of core shell studies, and catalyst pretreatment methods. Of concern to the committee is that many of these programs have been completed in late 2015 and it is unclear at this time which projects will continue (even within the recently announced FC-PAD consortium).

The portfolio of research projects also includes a spectrum of efforts from topics including gaining a fundamental understanding of critical issues related to the stability of Pt-containing catalysts, the interactions of ionomers with Pt nanoparticles, and the role of the ionomer in controlling oxygen transfer to the surface of the catalyst, an issue that becomes critical as the loading of Pt decreases. Efforts devoted to cathode performance at the system level, that take into account the effects of oxygen and proton transport as well as ORR kinetics as a function of the properties and interactions of the catalyst, support, and ionomer, and the interactions of catalyst-support-ionomer interaction with the membrane as a complete "system" are of significant value. Given the national laboratory facilities for doing atomic- and meso-scale characterization of components and complex systems, future research devoted to these subjects will add to the critical understanding of all classes of catalysts used in the electrode. The new consortia are expected to address these critically important activities.

A notable characteristic of a number of the preceding examples and many others reported at the DOE AMR meeting is their close relationship to technologies now believed to be under consideration for use by the OEMs, both foreign and domestic. This characteristic of the work is a consequence of requests from industrial partners in U.S. DRIVE for assistance in looking at specific issues related to in-house technology development and the desire to leverage high-end surface science and other spectroscopic characterization techniques and methods available at the national laboratories.

The discussion presented here highlights that communication of technology requirements and project results among all interested parties, Partnership related or not, must remain a priority. The DOE uses the AMR very effectively to accomplish this, including peer-review ratings of each project that are then used to assess future funding and support of a given project. A concern to the committee is that a number of the projects listed earlier at the national laboratories are ending or will soon be completed.

Significant Barriers and Issues That Need to Be Addressed

As in previous NRC reviews of the Partnership, cost and durability remain the critical barriers. With cost and durability still key issues, and knowing that the cost issue actually translates to new technical solutions, the committee understands the degree of effort and time required for the OEMs to meet the targets. With that said, the committee sees that appropriate DOE-funded projects are already in place. The FC-PAD presents an opportunity to address communication, coordination, program management, and teaming, and minimize duplication of effort and best use available resources, which will improve the chances of success in overcoming these challenges. Durability issues are predominantly materials-based and a function of system architecture and operating mode. New materials and solutions will result only from the continued focus on understanding the fundamentals of degradation mechanisms coupled with the continued investigations of carbon-free supports.

Water management in fuel cells has a dramatic impact on cost and durability as well as overall systems architecture. Though solutions exist in the current generation of vehicles, advanced humidification and water management schemes could reduce the complexity of the fuel cell if membranes could be developed that have a lower water requirement for proton transport. This will assist in simplifying system architecture (cost) and possibly allow for higher-temperature operation resulting in smaller and more efficient cooling systems. Electrode layer architecture and composition would also be positively impacted. The current funding level of new electrolytes for both the membrane and ionomer used in the electrode layer may not be sufficient and the DOE should consider rebalancing its portfolio to ensure that these areas are given proper emphasis.

Near-Term versus Long-Term Implications of DOE-Funded R&D

As has been pointed out previously, there is not a quantitative mechanism to assess the value and success of the DOE-funded efforts and the advancement of vehicle technologies via the Partnership. With that said, in recent years the DOE has funded approximately 56 projects, 42 related to fuel cell hardware, and 14 related to systems analysis, as reported at the 2015 AMR. Such projects address a spectrum of R&D efforts (e.g., at different technology readiness levels [TRLs]) that are predominantly long term (low TRLs) in nature—for example, new membranes, electrodes, catalysts, compressors, plates, operating performance under varying environmental conditions, to name a few. The longer-term projects align well with the objectives of the Partnership—that is, the projects are precompetitive, long term and high risk. Yet in any development effort, nearer-term learnings are natural fallouts and have the potential to bring immediate value to the OEMs as well as other parties, including the supply chain. It is important to distinguish between the two and maintain the proper perspective on the primary objective of such R&D efforts.

Appropriateness of Federal Funding

With the U.S. vehicle market moving toward near-zero-emission electric vehicles, the committee feels that it is appropriate that there is a portfolio of electric power generation sources under consideration, including those that utilize fuel cells. Therefore, the committee feels that federal funding is appropriate to facilitate technical solutions through support of R&D activities related to transportation. Though it is still premature, eventually a return of investment analysis[7] and an assessment of the benefits of the federally funded R&D by the DOE will be performed so as to assess the value of these programs to the United States.

Responses to Recommendations from the NRC Phase 4 Review

NRC Phase 4 Recommendation 3-3. The DOE should increase the efforts related to the development of new catalysts, membranes, and related membrane electrode assembly components for proton exchange membrane (PEM)-based fuel cells. The focus should be on materials, performance, durability, and, ultimately, on manufacturability.

Partnership Response. Within DOE's Fuel Cell Technologies Office (FCTO), a significant part of the Fuel Cell R&D portfolio is devoted to development of membrane electrode assemblies (MEAs) and related components for PEM-based fuel cells. It supports projects focused on the development of low-platinum group metal (PGM) and PGM-free catalysts, as well as membranes and MEA integration efforts targeting improved performance, enhanced durability, and decreased cost. Plans include expanding on these efforts, in line with DOE-supported cost analysis highlighting that catalysts dominate the PEM fuel cell cost projected at high volume manufacturing (500,000 units per year), whereas membranes dominate cost at low-volume production (1,000 units). As part of this continuing effort, FCTO released a Funding Opportunity Announcement, which explicitly solicited proposals on the topics of catalyst, membrane, and MEA component integration R&D. FCTO selected several projects either for immediate funding or as alternate projects to be initiated upon the availability of funds and subject to appropriations. For example, projects at 3M Corporation and Argonne National Laboratory focus on improving MEA performance via the integration of state of-the-art MEA components developed in other fuel cell R&D projects, while a Los Alamos National Laboratory project extends previous efforts to develop PGM-free cathode electrocatalysts.

Committee Assessment of Response to 3-3. Based on the data presented in Figure 3-5, the committee feels that the FCTO has been responsive to NRC Recommendation 3-3 of the Phase 4 report, especially for funding next-generation catalysts and electrodes. With respect to cell and other stack hardware and membrane electrode assemblies, funding levels saw an increase (2012-2014) in activity but then were essentially zeroed out in 2016. Last, funding for membranes and electrolytes have dropped significantly from the 2013 period to 2016, as previously shown in Figure 3-5. This is a concern of the committee.

[7] See Chapter 2, the section "The Role of the Federal Government."

NRC Phase 4 Recommendations 3-4 And S-4. The DOE should increase efforts for the cost reduction initiatives for fuel cells taking into account the entire system, including balance of plant. Emerging modeling capabilities should be used for sensitivity analysis and for guiding resource allocation to the areas that will have the greatest impact on performance, endurance, and cost at the system level.

Partnership Response. Modeling efforts have helped to identify the critical R&D areas and guide DOE decisions regarding resource allocation. For example, a significant part of DOE's fuel cell portfolio is devoted to development of membrane electrode assemblies (MEAs) for proton exchange membrane (PEM)-based fuel cells. The R&D portfolio supports projects focused on the development of low-platinum group metal (PGM) and PGM-free catalysts, as well as membranes and MEA integration efforts targeting improved performance, enhanced durability, and decreased cost. These efforts are in line with DOE-supported cost analysis highlighting the fact that catalysts dominate PEM fuel cell costs projected at high volume manufacturing (500,000 units per year), whereas membranes dominate cost at low-volume production (1,000 units). In addition, a recent DOE competitive funding opportunity led to the selection of Eaton Corporation, which will develop an air management system with an integrated expander for the fuel cell system balance of plant.

Committee Assessment of Response to 3-4 and S-4. The committee feels that this area has been addressed within this review period, especially with the systems analysis group, yet durability and cost (from a technical perspective) solutions have not yet emerged from such efforts. The committee questions the value of performing low-volume (100,000 units per year) manufacturing analyses.

NRC Phase 4 Recommendation 3-5. Either in coordination with other organizations, such as the Office of Basic Energy Sciences or DOE ARPA-E, or directly, DOE should consider supporting new and innovative alternative fuel cell concepts.

Partnership Response. DOE supports this recommendation, and the Fuel Cell Technologies Office (FCTO) does support R&D projects addressing longer-term, innovative fuel cell concepts, such as reversible fuel cells, novel fuel cell structures, and alkaline membrane fuel cells. A recent FCTO R&D Funding Opportunity Announcement specifically included an innovative concepts topic soliciting novel ideas, with a primary thrust on new materials, new architectures, or new modes of operation for fuel cells. Furthermore, FCTO has been actively coordinating with DOE's Office of Basic Energy Sciences and ARPA-E to further enhance collaboration supporting innovative R&D. Novel alkaline-exchange membrane R&D at Los Alamos National Laboratory is an example of demonstrated leveraged support across agencies.

Committee Assessment of Response to 3-5. The committee agrees that there has been good coordination of fuel cell development activities among the different offices as determined from the report outs at the Annual Merit Review meetings. The committee also agrees that innovative topics, as recommended in the NRC Phase 4 report, have been supported as cited in the Partnership's response.

NRC Phase 4 Recommendation 3-6. U.S. DRIVE should encourage projects that address the use of real-time, in situ electro-analytical quality-control methods to assess membrane and electrode performance characteristics during the continuous manufacturing web-based process.

> ***Partnership Response.*** The Partnership, including the U.S. Department of Energy (DOE), agrees with this recommendation. In DOE's Fuel Cell Technologies Office, the Manufacturing subprogram currently supports a project at the National Renewable Energy Laboratory (NREL) to develop in situ quality control methods for membrane electrode assemblies (MEAs) and to correlate defects introduced during manufacturing with fuel cell performance. The U.S. DRIVE Fuel Cell Technical Team supports further expanding these efforts to improve the MEA manufacturing process and encourages efforts to understand structure/property/performance relationships, with the next step being the development of quality control methodology to identify and control critical characteristics in the MEA manufacturing process.

Committee Assessment of Response to 3-6. The committee feels the response has been appropriate.

Findings and Recommendations

The overall assessment based on the review is that the U.S.-based OEMs, with significant input from the Partnership, although in different states of development, have advanced fuel cell technology to the point where at least one U.S. DRIVE Partnership OEM (General Motors) is anticipating a rollout of its HFCV in 2020. With cost and durability challenges remaining, the DOE appropriately, and in response to the Partnership's mission, continues to support longer term, precompetitive R&D projects. Furthermore, the DOE is adding additional activities at the national laboratories with the creation of consortia which will help focus and coordinate the R&D.

Finding 3-16. Since the NRC Phase 4 review, Toyota, Hyundai, and Honda have made available within the United States a limited number of fuel cell vehicle sales or leases to the general public. U.S.-based OEMs, with significant input from the Partnership, although in different states of development, have advanced fuel cell technology to the point where at least one U.S. DRIVE Partnership OEM (General Motors) is anticipating a rollout of its fuel cell vehicle in 2020. The development and deployment of roadworthy fuel cell vehicles is a major accomplishment and one that will help to identify remaining technical, cost, manufacturing, and infrastructure challenges. Though the cars are still in the late stages of development, the fact that the cars have advanced to this point is due in part to R&D coordination by the Partnership and its prior organizations, as well as from decades of funding of pertinent research projects by the DOE and Partnership members.

Finding 3-17. With the U.S. OEMs in different states of fuel cell vehicle development, and with competitive dynamics emerging, selected Partnership (fuel cell) goals and targets are relevant to only some of the OEM members (e.g., Pt loadings). Furthermore, it appears that there is a fine line between what might be considered near- and long-term projects based on the state of development of a given OEM's technology.

Recommendation 3-5. The Partnership should evaluate projects for their near-term or long-term potential impact and assign technology readiness levels to them. The Partnership should continually assess its process for prioritizing projects and should continue to address the longer-term, precompetitive (lower technology readiness level) objectives and should update and set longer-term targets.

Finding 3-18. DOE is in the process of forming the Fuel Cell Performance and Durability consortium and the ElectroCAT consortium that will focus on fuel cell electrode durability science and electrochemical engineering, results of which should positively impact fuel cell performance and lifetime (durability).

Recommendation 3-6. The newly formed Fuel Cell Consortium for Performance and Durability and Electrocatalysis consortium should have an active oversight committee comprised of a panel of non-national laboratory experts, including representation from the original equipment manufacturers, to ensure that relevant problems are being addressed. Department of Energy-funded projects should focus on developing understandings of the fundamental issues relating to reduction in platinum loading to fuel cell durability.

Finding 3-19. Significant funding resources continue to be directed at carbon-supported Pt-based electrodes. Projects related to non-PGMs are in early stages of emphasis.

Recommendation 3-7. The Department of Energy should increase its focus on non-carbon-supported platinum catalysts and noncarbon catalyst support materials. In addition, the Department of Energy should increase the effort on developing non-platinum-group-metal catalysts and electrodes.

ONBOARD HYDROGEN STORAGE

Background

The mission of the hydrogen storage technical team (HSTT) is to "accelerate research and innovation that will lead to commercially viable hydrogen-storage technologies that meet the U.S. DRIVE Partnership goals" (U.S. DRIVE, 2013g). HSTT is one of two joint vehicle-fuel technical teams with a line to both the Vehicle Operations Group and the Fuel Operations Group within the U.S. DRIVE Partnership. The HSTT members are from the three car companies (GM, Ford, and FCA) plus representatives from the DOE and the national laboratories. The technical team has one associate member from a university. They have key interactions with several other technical teams, namely fuel cells, hydrogen production, hydrogen delivery, fuel pathway integration, hydrogen codes and standards, and vehicle systems and analysis (see Figure 2-1 in Chapter 2). These interactions involve such issues as fuel quality, materials compatibility, storage capacity, and safety.

Vehicle driving range and fueling time are important customer attributes for fuel cell vehicles. The objective is to achieve a driving range of at least 300 miles for a full range of light-duty vehicles and at the same time meet performance, packaging, cost, rapid fueling time, and safety requirements.

A dual approach is being taken to address these goals. The near-term technology focus is on the 700-bar compressed hydrogen system. The longer-term approach is cold/cryo-compressed hydrogen storage systems and materials-based storage systems. Early in the program (2005-2010) materials R&D was conducted through three hydrogen storage centers of excellence, each managed by one of the national laboratories and each one focused on one of three technology areas: metal hydrides (led by SNL), chemical hydrogen storage (led by LANL), and sorbents (led by the NREL). These centers facilitated communication among research groups, developed best practices for research and a mechanism to evaluate progress, and were succeeded by the Hydrogen Storage Engineering Center of Excellence (HSECoE) led by the Savannah River National Laboratory, which ran from February 2009 through December 2015. This center was focused on engineering analysis and system integration modeling. The design and build of subscale prototype systems was used to validate the models and improve designs, and as a predictive tool to define material properties to meet the DOE system targets. These centers of excellence are now closed.

Recently a new consortium has been formed called the Hydrogen Materials—Advanced Research Consortium (HyMARC) which couples basic and applied R&D with advanced computational materials design. The approach taken is to have a core laboratory team comprised of SNL (lead), LLNL, and LBNL that conducts fundamental science and develops resources for the R&D community. The other component is comprised of individual projects selected through Funding Opportunity Announcements/Lab calls. Material characterization and validation resources are at NREL (lead), LBNL, Pacific Northwest National Laboratory (PNNL), and the National Institute for Standards and Technology. The long-term focus will be on multiscale simulations. The goal is to have an improved and faster approach to identify promising materials. The laboratory teams make use of DOE through the Office of Science (the Basic Energy Sciences [BES] and Advanced Scientific Computing Research [ASCR]) user facilities.

The U.S. DRIVE HSTT roadmap was last issued in June 2013 (U.S. DRIVE, 2013g). This roadmap contains detailed information on key issues and challenges, technical targets, technical barriers, and R&D strategies to achieve the technical targets.

DOE funding for the onboard hydrogen storage activities are bundled in the hydrogen fuel R&D line item, which includes hydrogen production and delivery R&D and hydrogen storage R&D. The DOE funding for hydrogen fuel R&D was \$41.05 million for FY 2016 (Satyapal, 2016).

The DOE hydrogen storage projects as reported at the annual AMR reviews comprise a significant part of the preceding funding, which has remained constant at $15.6 million from 2014 through 2016 annually (Stetson, 2014, 2015, 2016). The requirement that projects be fully funded at the start of the project (a recent change) makes year-to-year comparisons difficult.

The current status of onboard hydrogen storage technologies is shown in Table 3-8. The storage targets are based on 5.6 kg of usable hydrogen and an 80 kW net power fuel cell. This status can be compared with the storage targets established for 2020 and the U.S. DOE ultimate targets. The technical targets have been revised since the NRC Phase 4 report was issued (NRC, 2013). Progress has been made in both near-term and long-term technologies. No-go decisions made by the DOE have provided for a focused effort on new and promising technologies.

Assessment of Progress and Key Achievements

The projects that address onboard hydrogen storage include both near-term and long-term technologies. The near-term technologies include issues related to cost and performance of the carbon fiber-based hydrogen storage vessels. The longer-term technologies include technologies related to the cryo/compressed storage vessels and storage materials with emphasis on meeting the U.S. DRIVE targets. The U.S. DOE Fuel Cell Technologies Office provided the committee with

TABLE 3-8 Current Status of Hydrogen Storage Technologies

Targets/Projected Systems	Gravimetric kWh/kg (kg H_2/kg system)	Volumetric kWh/L (kg H_2/L system)	Costs (2007\$)[a] \$/kWh (\$/kg H_2)
2020 Storage targets	1.8 (0.055)	1.3 (0.040)	10 (333)
Ultimate storage targets	2.5 (0.075)	2.3 (0.070)	8 (266)
700 bar compressed (Type IV, single tank)	1.4 (0.044)	0.8 (0.024)	15 (500)
Metal hydride ($NaAlH_2$/Ti)	0.4 (0.012)	0.4 (0.012)	43 (1,432)
Sorbent (MOF-5, 100 bar, MATI, LN2 cooling)	1.3 (0.04)	0.7 (0.020)	16 (533)
Chemical hydrogen storage (AB-50 wt%)	1.4 (0.043)	1.3 (0.040)	17 (566)

[a] Projected to 500,000 units/year.
SOURCE: Stetson and Veenstra (2016).

a list of projects related to onboard hydrogen storage that it funded in recent years (Table 3-9). The committee sorted these by topic for clarity.

The carbon fiber resin and matrix represent more than one-half of the cost of the compressed storage system for the carbon-fiber-based hydrogen storage tanks. Progress has been made through the use of low-cost textile-based polyacryrlonitrile (PAN) precursors to reduce the cost of carbon fiber. The PAN precursor represents an 18 percent reduction in carbon fiber cost which brings the carbon fiber price down to $10.6/lb from a 2013 baseline cost of $13/lb. Both costs assume an annual production of 25,000 tons. Progress has also been made on reducing system cost based on balance of plant component integration. There is a strong synergy between the carbon fiber interests of the onboard hydrogen storage technical team and the interest in lightweight materials of the materials technical team.

Although the progress in cost reduction of compressed hydrogen storage tanks quoted by the DOE is promising, it should be noted that the Toyota Corporation has published a recent technical paper (Yamashita et al., 2015) in which it claims that its latest compressed hydrogen storage system is lighter in weight (15 percent reduction) due to the use of advanced carbon fiber, new winding designs, and simplified balance of plant components. This tank design, currently used in limited production vehicles, is reduced in cost compared to the previous 2008 tank used in prototype fuel cell vehicles. It has also been validated for drop and fire safety regulations. Additional details on the absolute numbers for weight and cost are needed in order to fully assess these claims.

There are numerous reasons for the DOE to invest in R&D programs directed at improving carbon and glass fiber strength and reducing cost for applications not only in the automotive industry but in other industries requiring improved efficiency. However, improving the design of the compressed gas storage tanks (project ST101) and reengineering the balance of plant (projects ST005 and ST113) is an effort that is currently being conducted by both OEMs and suppliers, making it unnecessary for the DOE to conduct programs directed at these objectives. Allowing the engineering development to be done by industry would allow the DOE to focus on developing solid-state storage materials and advanced composites.

Cold/cryogenic compressed hydrogen storage is being pursued as a long-term option for compressed hydrogen storage. Potential advantages are lower operating pressure, higher energy density, less carbon fiber for tank construction, and lower system cost depending on the approach taken (see Figure 3-11). The two approaches are as follows: The cryogenic compressed system has a target volumetric capacity of >50 g/L hydrogen and >9 weight percent (wt%) H_2 system capacity at 700 bar and 40 K in contrast to 24 g/L for the 700-bar system at 295 K. The same amount of usable hydrogen could thus be stored in about 50 percent less volume (a smaller tank and at lower cost). The cold compressed hydrogen storage system has a target gravimetric capacity of 8-9 wt% H_2 at

TABLE 3-9 U.S. Department of Energy Fuel Cell Technologies and Vehicle Technologies Offices Active Project List Related to Onboard Hydrogen Storage That Supports U.S. DRIVE Targets

Project	Organization	Presentation
700 Bar Compressed and Cold/Cryo Compressed Storage System Projects		
ST111	LLNL	Thermomechanical Cycling of Thin-Liner, High-Fiber-Fraction Cryogenic Pressure Vessels Rapidly Refueled by Liquid Hydrogen Pump to 700 bar
ST114	Materia	Next-Generation Hydrogen Storage Vessels Enabled by Carbon Fiber Infusion with a Low-Viscosity, High-Toughness Resin System
ST093	ORNL	Melt-Processable PAN Precursor for High-Strength, Low-Cost Carbon Fibers
ST005	PNNL	Systems Engineering of Chemical Hydrogen, Pressure Vessel, and Balance of Plant for Onboard Hydrogen Storage
ST101	PNNL	Enhanced Materials and Design Parameters for Reducing the Cost of Hydrogen Storage Tanks
ST115	PPG Industries	Achieving Hydrogen Storage Goals through High-Strength Fiber Glass
ST113	SNL	Innovative Development, Selection, and Testing to Reduce Cost and Weight of Materials for Balance-of-Plant Components
ST100	Strategic Analysis	Hydrogen Storage Cost Analysis
ST110	Composite Technology Development, Inc.	Optimizing the Cost and Performance of Composite Cylinders for Hydrogen Using a Graded Construction
ST126	Center for Transportation and the Environment	Conformable Hydrogen Storage Coil Reservoir
HSECoE and Related Projects		
ST0001	ANL	System-Level Analysis of Hydrogen Storage Options
ST008	NREL	System Design, Analysis, and Modeling for Hydrogen Storage Systems
ST004	Savannah River National Laboratory	Hydrogen Storage Engineering Center of Excellence
Materials-Based Storage Projects		
ST118	LLNL	Improving the Kinetics and Thermodynamics of $Mg(BH_4)_2$ for Hydrogen Storage
ST122	University of Michigan	Hydrogen Adsorbents with High Volumetric Density: New Materials and System Projections
ST119	Ames Laboratory	High-Capacity Hydrogen Storage System via Mechanochemistry

TABLE 3-9 Continued

Project	Organization	Presentation
ST121	Texas A&M University	High-Capacity and Low-Cost Hydrogen-Storage Sorbents for Automotive Applications
ST120	California Institute of Technology	Design and Synthesis of Materials with High Capacities for Hydrogen Physisorption
ST014	NREL	Hydrogen Sorbent Measurement Qualification and Characterization

NOTE: Acronyms defined in Appendix D.
SOURCE: Project numbers are from the Department of Energy's 2016 Annual Merit Review, see Cooper (2016a).

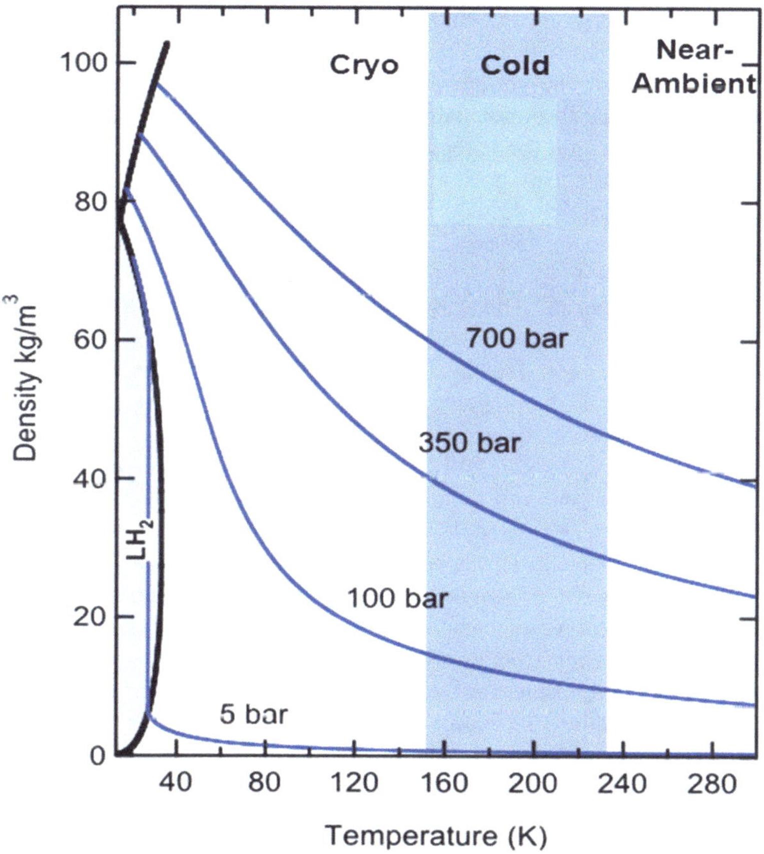

FIGURE 3-11 Analysis showing the impact of cryo-compressed hydrogen storage.
NOTE: Page 19 (right side only) of presentation.
SOURCE: Ned Stetson and Mike Veenstra, Hydrogen Storage Technical Team Presentation to the NRC Committee on the Review of the U.S. Drive Partnership, April 19, 2016.

500 bar and 200 K compared with 4.4 wt% H_2 for the 700-bar system at 295 K. The same amount of usable hydrogen could thus be stored with 50 percent less system mass because of the lower pressure. During the past year the focus of the cryo/compressed project included pressure vessel design, fabrication, and testing, including electricity consumption on refueling. The energy efficiency of the total system operation (including the energy required for hydrogen liquefaction) is a major factor that needs to continue to be addressed.[8]

Materials-based storage continues to hold promise in the long term; however, many of these efforts have been redirected as they will not meet the 2020 targets. Work on the metal hydride system (e.g., sodium alanate) was discontinued based on HSECoE model projections. The materials under study lacked suitable thermodynamic, kinetic, and gravimetric properties. Progress was reported on an electrochemical process to reduce alane synthesis cost. Work on the chemical hydrogen storage system (e.g., ammonia borane) was discontinued due to lack of liquid phase materials with suitable regeneration processes. The committee supports and commends these programmatic decisions, which were based on model projections and clear insurmountable barriers.

Work continues on the sorbent system (e.g., MOF-5) for cold 80-160 K, 100-bar storage. Work in progress includes heat exchanger concepts and prototype testing. The HSECoE has developed system models that can be applied to materials, storage tank mass, and cost. The system models that capture the hydrogen storage systems, the fuel cell system, and model at the vehicle level are available on the HSECoE website.[9] The work scope of several projects is supported by the HSECoE.

The Highlights of Technical Accomplishments Report (U.S. DRIVE, 2015) cites two areas of progress in the area of hydrogen storage. These are (1) Adsorption-Based Hydrogen Storage System Validated and (2) Lower-Cost Hydrogen Storage System. In (1) HSECoE subscale prototypes were used to validate the models and improve their predictive capability. Two prototype systems were used to characterize materials performance, internal heat exchange capability, and thermal management. This approach was applied successfully to confirm the robustness of the (MOF-5) adsorbent material. The second accomplishment cited (2) is the lower cost PAN precursor-based textile processing mentioned above.

Onboard hydrogen storage is cited in the DOE program reviews to be of interest to several technical teams, even beyond HSTT. For example, the materials technical team is addressing hydrogen storage tank materials, the hydrogen codes and standards technical team is addressing safe deployment, and the hydrogen delivery technical team shares needs for hydrogen storage. Collaboration among

[8] Estimates of the energy required for liquefaction vary and can be found in the following DOE document: https://www.hydrogen.energy.gov/pdfs/9013_energy_requirements_for_hydrogen_gas_compression.pdf.

[9] The website for the Hydrogen Storage Engineering Center of Excellence is http://hsecoe.org, accessed September 12, 2016.

the technical teams is increasing as the various technologies and commercialization needs mature.

HSTT participates in and monitors international hydrogen storage activities. These activities include the International Energy Agency—Hydrogen Implementation Agreement (Task 32) and international conferences.

Significant Barriers and Issues That Need to Be Addressed

While significant progress has been made toward the 2020 targets, the major barriers that need to be addressed are gravimetric density, volumetric density, and system cost for the projected 700-bar-type system. Practical limitations prevent the 700-bar system from meeting all of the onboard hydrogen storage targets. (The use of larger or multiple tanks in order to carry more hydrogen on board the vehicle is viewed as a solution for some applications.) Targets must be met across a wide range of vehicle platforms and especially for those platforms where the benefits are greatest. (Yamashita et al. [2015] cites gravimetric density exceeding the 2020 target.)

Significant barriers that are the focus of ongoing R&D are specific to each technology area.

- The near-term 700-bar compressed hydrogen system requires lower-cost carbon fiber, improved composites, innovative winding designs, conformable designs, and lower-cost balance of plant.
- The cold/cryo-compressed hydrogen storage system requires advanced insulation, improved dormancy, and composite development.
- The materials storage systems (metal hydrides, sorbents, and chemical H_2 storage) all need lower system cost and higher material capacity. Fill time and onboard efficiency is an issue for the metal hydrides (e.g., $NaAlH_4$). Dormancy and well-to-power plant efficiency is an issue for the sorbents (e.g., MOF-5). The chemical hydrogen storage systems (e.g., NH_3BH_3) need lower-cost off-board regeneration.

The need to overcome these barriers does not preclude vehicle introduction because advancements in overall power train system efficiency and vehicle weight reduction can make up for a shortfall in hydrogen storage capacity.

Responses to Phase 4 Recommendations from Phase 4 Review

NRC Phase 4 Recommendation 3-7. The U.S. DRIVE Partnership should re-examine high-pressure compressed gas storage and reach a consensus as to whether this is a long-term solution or just a transition technology. Short-term and medium-term performance targets should be developed specifically for compressed tanks because such tanks are expected to be used at least on the first generation of hydrogen fuel cell vehicles. Then there should be long-term general materials targets that basic research can use for benchmarking.

Partnership Response. Vehicle manufacturers have demonstrated that 350 and 700 bar compressed hydrogen storage systems can provide a sufficient driving range and performance to meet customer expectations on some vehicle platforms and therefore provide a pathway for initial commercialization. The Partnership developed its on-board hydrogen storage system targets based on projected requirements to meet customer performance expectations across the range of light-duty vehicle platforms; they are not technology specific. Partners will continue their analysis in an effort to determine an optimum balance between onboard performance and costs, and offboard efficiency and costs. Additionally, advanced hydrogen storage technologies will continue to be developed and evaluated for their potential to meet all of the onboard storage requirements and compared against high-pressure storage. Evaluations will include impact on hydrogen refueling infrastructure and potential horizon for introduction. The Partnership agrees that a complete system analysis of materials-based technologies would be valuable to translate the material properties required to meet system targets and for use in guiding materials development research efforts. The U.S. Department of Energy's Hydrogen Storage Engineering Center of Excellence is currently carrying out this analysis.

Committee Assessment of Response to 3-7. The committee considers that the actions taken by the Partnership regarding high-pressure onboard hydrogen storage and systems analysis of materials-based technologies to be fully responsive to the recommendation.

NRC Phase 4 Recommendation 3-8. The U.S. DRIVE Partnership should investigate the relationship between the onboard hydrogen storage tank pressure and the hydrogen infrastructure so that trade-offs can be worked out.

Partnership Response. The Partnership agrees with this recommendation. The Hydrogen Storage Technical Team is working with the Fuel Pathway Integration and Hydrogen Delivery Technical Teams to better understand the cost and efficiency implications of onboard storage pressure on hydrogen refueling infrastructure. The U.S. Department of Energy plans to continue these analyses to assess the cost and performance impacts associated with varying the onboard storage pressure. The Technical Teams are also looking at additional variables, such as temperature and thermal management for materials-based storage systems. These trade-offs can be considered, but must be balanced with the customer value of driving range, which is often a shortcoming of fuel cell vehicles in comparison to conventional vehicles.

Committee Assessment of Response to 3-8. The committee supports the involvement of the other technical teams in addressing the relationship between the onboard hydrogen storage tank pressure and the hydrogen infrastructure and concludes that it is fully responsive to the recommendation.

NRC Phase 4 Recommendation 3-9. The U.S. DRIVE Partnership should consider joint programs with the U.S. Department of Defense and the National Aeronautics and Space Administration, which undoubtedly have similar goals for lower-cost aerospace quality carbon fibers. Work with the newly constructed ORNL Carbon Fiber Technology Facility should also be explored.

Partnership Response. The Partnership supports this recommendation. The U.S. Department of Energy (DOE) has and will continue to pursue partnerships both internally and with other government agencies to help advance its mission, leverage resources, and eliminate duplication of efforts. Examples include a February 2011 workshop on carbon fiber composite pressure vessels, including carbon fibers, which included various industry and government stakeholders, and a joint project involving DOE's Fuel Cell Technologies and Vehicle Technologies Offices, working with ORNL to develop lower cost precursor material for higher strength carbon fiber. This work, initiated in 2010, will be expanded to include pilot scale activities at ORNL's Carbon Fiber Technology Facility when appropriate.

Committee Assessment of Response to 3-9. The committee believes that the activities regarding the partnerships with other government agencies and DOE offices regarding lower-cost aerospace quality carbon fibers are responsive to the recommendation.

NRC Phase 4 Recommendation 3-10. The U.S. DRIVE Partnership should demonstrate the safety of lower-cost, lighter-weight compressed hydrogen tanks with a rigorous testing program, for example, by statistically demonstrating stress rupture toughness, fatigue life, and fire safety. In implementing such an activity, it should consider co-funding the related tests proposed by the NASA White Sands facility.

Partnership Response. The Partnership agrees with this recommendation. In addition to carrying out development efforts on lighter-weight and lower-cost materials of construction and improved designs for compressed hydrogen tanks, the U.S. Department of Energy (DOE) and U.S. DRIVE industry partners collaborate with codes and standards development organizations, regulatory bodies (both domestic and international), and other stakeholders to develop rigorous and comprehensive design and testing protocols to ensure the safe design and manufacture of compressed hydrogen tanks. These efforts are carried out in collaboration with numerous stakeholders that include various U.S. and international government agencies as well as commercial entities. For example, DOE funded a round-robin tank testing on Type 4 tanks to determine the proper test measurement protocols, with the U.S. testing taking place at the NASA White Sands facility.

Committee Assessment of Response to 3-10. The committee considers that the actions taken regarding the development of rigorous and comprehensive design and testing protocols to ensure the safe design and manufacture of compressed hydrogen tanks in conjunction with industry and government stakeholders are responsive to the recommendation.

NRC Phase 4 Recommendations 3-11 and S-5. The DOE (e.g., the Office of Basic Energy Sciences, the Office of Energy Efficiency and Renewable Energy, the Advanced Research Projects Agency-Energy) should initiate a new program that builds on the excellent progress made to date and expands into fundamentally new hydrogen storage research areas. A critical assessment of prospects for, and barriers to, advanced storage techniques and concepts should form the first part of this initiative.

Partnership Response. Each U.S. Department of Energy (DOE) office maintains an R&D project portfolio to achieve its specific mission objectives. The offices communicate and collaborate on common and overlapping areas of interests. Current common areas of activity and interest include low-cost, light-weight materials of construction (such as carbon fiber); highly porous gas sorbents; and low-cost conformable storage tanks. DOE may consider a new interoffice initiative on hydrogen storage as resources allow.

Committee Assessment of Response to 3-11 and S-5. The committee believes that the coordination of activities within the DOE with BES and ARPA-E with respect to hydrogen storage is a good first step and looks forward to a more robust interoffice initiative to pursue this critical task.

Appropriate Federal Role

The DOE hydrogen storage materials projects are an appropriate role for federal funding given the benefits of fuel cells including high efficiency and low greenhouse gas emissions when the hydrogen fuel is made from renewable energy sources. Advanced hydrogen storage technologies need to be developed that meet system cost targets and onboard storage targets for volumetric density and gravimetric density. DOE-funded precompetitive R&D can lead to both new fundamental understanding and new technologies in order to meet the cost and capacity targets. The national laboratories, university laboratories, and selected industry laboratories are well equipped to contribute to this effort. Tank design and development is being conducted at this time by OEMs and suppliers and is being implemented in production vehicles. As such these are now competitive activities.

Findings and Recommendations

Finding 3-20. The 700-bar hydrogen storage tank will be the technology used for the immediate future and this technology, although continuing to evolve, will be used in production vehicles. Materials-based storage has been extensively researched in a well-organized DOE program. Materials that still hold promise and new discoveries could have a high hurdle if they are to displace technology in place. The new DOE Consortium HyMARC will address scientific gaps blocking the advancement of solid-state storage materials. The HSTT roadmap was last updated June 2013 and significant progress and changes have been made in recent years, including technical targets.

Recommendation 3-8. The Department of Energy should not be involved in design, development, or testing of production tank components, but the following actions are needed:

a. The mission for the hydrogen storage technical team should be made more explicit in terms of meeting critical needs and a plan should be developed for reaching these goals and targets.
b. The hydrogen storage technical team roadmap should be updated by the Partnership again in light of recent developments.

c. The management structure for the Hydrogen Materials—Advanced Research Consortium should provide for coordination, prevention of duplication, and decision-making authority over the consortium projects.

Finding 3-21. All the goals for onboard hydrogen storage have not been met, and basic scientific research has not produced an easy solution to date. Yet, onboard hydrogen storage is an issue for several technical teams and working groups beyond the hydrogen storage technical team—for example, the materials technical team, the fuel cells technical team, the hydrogen codes and standards technical team, and the hydrogen delivery technical team. As the technologies continue to mature, the need to merge activities can be expected to increase because vehicle performance parameters might be achieved through a wider range of options than gravimetric and volumetric hydrogen storage density alone.

Recommendation 3-9. The hydrogen storage technical team should increasingly work with the other technical teams even beyond those areas where overlap currently exists.

HYDROGEN PRODUCTION, DELIVERY, AND DISPENSING

Background and Introduction

The supply and refueling logistics, associated technologies, as well as the cost of hydrogen at the pump are critical elements of the transformation process in which HFCVs displace fossil fuel-powered internal combustion engines. At the time of the writing of this report, there were approximately 25 fueling stations in the United States, mostly in California, based on a number of technologies. In selected areas, mobile refuelers are used by some OEMs. Up until now only a very small number of HFCVs have been on the road, yet with the projections by the OEMs, this number will increase rapidly in the years to come. Though in its infancy, the hydrogen refueling infrastructure must be addressed in its entirety, from generation to delivery, as well as the fueling station itself and ancillary activities such as safety, codes, and standards.

This section reviews the budgets, goals, targets, and accomplishments of the U.S. DRIVE Partnership related to hydrogen production, delivery, and dispensing. A discussion of issues and barriers and the role of the government in overcoming such barriers is also included along with committee recommendations.

As mentioned in the introduction, one of the focus areas of the U.S. DRIVE Partnership is to provide a coordination mechanism by which OEMs interested in HFCVs can communicate precompetitive technical issues and target specifications to its members. Among the operations groups defined in the U.S. DRIVE Partnership Plan, the Fuel Operations Group (see Chapter 2) to date has almost entirely been focused on hydrogen—its production, delivery to fueling stations, dispensing, fuel pathway integration, and onboard storage. With the Partnership guidance and based on other inputs, the DOE has been funding a number of production and delivery projects as a means to begin the learning process for when the fuel cell vehicles start to appear in the market. However, while the U.S. DRIVE industry partners appear to be sufficiently engaged in vehicle-based developments relevant to the Partnership's goals and targets, their involvement in activities related to the hydrogen supply chain development, including production, delivery, and dispensing, appears to be limited, especially since the NRC Phase 4 review.

Introduction of HFCVs in the market and presence of an adequate hydrogen infrastructure are often seen as a "chicken-and-egg" problem—that is, who will invest in hydrogen fueling infrastructure if there are too few HFCVs on the road and, conversely, who will offer HFCVs if there are a limited number of hydrogen fueling stations. The DOE program goal relevant to U.S. DRIVE is "to enable a commitment by automakers *no later than year 2015* to offer safe, affordable, and technically viable hydrogen fuel cell vehicles in the mass consumer market" (Satyapal, 2016). This does not seem to have been fulfilled by U.S. DRIVE auto industry partners, except perhaps GM, although Toyota, Hyundai, and Honda are often quoted by DOE as examples of automakers with commercial plans. Hyundai and Toyota are already offering commercial HFCVs in the United States. Toyota and Honda are also engaged in refueling activities. Toyota has already begun a rollout of its fuel cell vehicles; Honda has begun leasing vehicles (Clarity) and together with GM are targeting to introduce their vehicles by 2020 (Greimel, 2016). U.S. auto manufacturers have not publicly announced similar plans as yet, although GM so far has the most number of HFCVs (119 Equinox) with more than 3 million miles of on-road driving experience (General Motors, 2014, 2016).

As pointed out before, of the total DOE EERE annual budget for hydrogen and fuel cells, which has been more or less stable over the last 3 to 4 years at about \$100 million (± 5 percent), roughly \$35 million is spent on hydrogen production, storage, and delivery, which is about the same as that for fuel cells (Sarkar, 2015).[10] Furthermore, production and delivery budgets, including those for FY 2016, are approximately \$12.5 million each. There is an additional budget of \$7 million each for technology validation activities and for Safety, Codes, and Standards (SCS) for hydrogen as a fuel (Satyapal, 2016). Much of the hydrogen R&D effort is appropriately directed to reducing the cost of dispensed hydrogen by focusing on key steps in the value chain. However, as shown by the fuel pathway integration analysis results with the current technologies, it is challenging to meet the cost targets by the prescribed dates (Joseck and Verduzco, 2016). More efforts will need to be focused on new concepts and next-generation technologies.

Since hydrogen is the lightest gas it has a very low volumetric energy density. It therefore requires compression to high pressure (e.g., 700 bar) or liquefaction (20 K) to store and transport as well as store on the vehicle before use. Thus,

[10] Also see Peterson and Farmer (2016).

hydrogen supply to an HFCV involves a number of steps from production to the vehicle tank. One example of the supply chain for gaseous hydrogen is shown in the schematic in Figure 3-12.

Hydrogen production from fossil-based feedstocks, especially by reforming natural gas, is commercially well established. In the reforming process, all the carbon in natural gas is ultimately converted to CO_2 and released to the atmosphere. To avoid emitting the CO_2 in this pathway would require capture and storage and/or utilization of the CO_2 (CCSU). This of course would add cost and make the hydrogen more expensive. The other current pathway for hydrogen production is through the electrolysis of water; the associated GHG emissions in this case are dependent on the source of electricity. Further reduction in GHG emissions requires use of renewable resources, such as a biomass feedstock and/or use of wind- and solar-based electricity. Other long-term approaches also being explored include solar thermochemical, photoelectrochemical, and microbial biomass conversion.

While the Partnership is focused on the long-term, precompetitive R&D, it should be borne in mind that the near-term needs of hydrogen are met by existing, established production technologies, primarily using natural gas reforming, that meet the DOE production cost target of \$2/kg H_2. Purification will likely be a small part of the cost. However, current delivery and station costs estimated at \$11-14/kg H_2 (Miller, 2016) are high, and are mainly responsible for the high

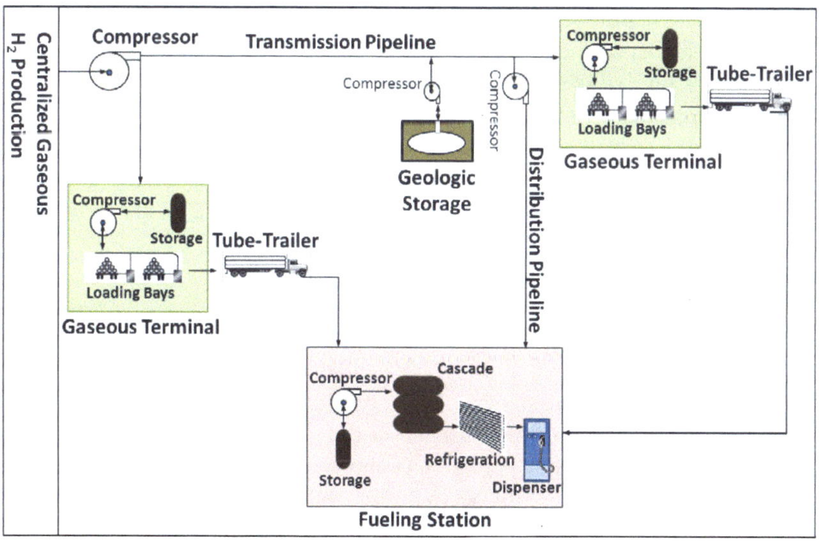

FIGURE 3-12 Schematic of production and delivery of hydrogen to a fueling station. SOURCE: U.S. DRIVE (2013f).

total cost of dispensed hydrogen of \$13-16/kg H_2 compared to the 2020 DOE cost target for untaxed dispensed hydrogen of \$4/kg. High-volume usage is expected to bring these costs down to the \$5-7.5/kg H_2 range. But the challenges to meet the ultimate cost targets are related to the cost of "renewable" hydrogen production and more importantly the cost of delivery and dispensing (station cost). Therefore, priorities between long-term production R&D and near-term developments related to hydrogen delivery and dispensing are often difficult to balance, given limited funding available. DOE-funded programs have made progress to improve technologies for high-pressure gaseous hydrogen storage, delivery, and dispensing by addressing and optimizing the value chain. Another related area that is getting attention from a fueling station perspective is SCS. However, it is not clear if the costs associated with the SCS requirements (e.g., for additional equipment and insurance) are adequately accounted for in the cost estimates for dispensed hydrogen.

Critically important are infrastructure build-out activities necessary to support a commercialization pathway for HFCVs and to get real-world experience with these vehicles, which will help identify further R&D needs. GM, for example, which is actively developing fuel cells and fuel cell vehicles, expects states to incentivize fueling station build-out and considers existence of a viable fueling infrastructure a prerequisite to vehicle introduction.[11] Thus, due to uncertainty of demand and volume, hydrogen infrastructure is highly dependent on federal and state funding; automotive and energy industry investment toward these efforts is minimal. Without adequate funding/investment in hydrogen infrastructure, commercialization of HFCVs will be pushed even further back beyond 2020. The lack of a robust plan for a viable national hydrogen infrastructure is a significant deterrent to the deployment of HFCVs.

Given the high risks and costs associated with initial introduction of HFCVs, for-profit companies are reluctant to participate on their own, and therefore federal and state roles in funding such activities are critically important and appropriate. The cost challenges combined with practical challenges of installing a hydrogen infrastructure imply that a paradigm shift is required if HFCVs are to become a realistic option for consumers.

Current Status of U.S. DRIVE Goals and Targets

With the limited number of HFCVs on the road today, resulting in low demand for hydrogen, economy of scale advantage is not realized. As a consequence, at present, the cost of hydrogen at the pump is very high (\$13 to \$16/gal gasoline equivalent (GGE) or kg H_2) compared to the DOE cost target of less than \$4/kg H_2 necessary for HFCVs to be competitive with other options, specifically HEVs (Sutherland and Joseck, 2015). Of the total cost, the current produc-

[11] The committee gathered this information during a visit to General Motors on June 21, 2016.

tion cost is relatively low, since most of the hydrogen is sourced from existing large production facilities using natural gas as the feedstock; a major portion of the cost is for delivery and dispensing. While U.S. DRIVE has a cost target for onboard hydrogen storage, it does not have a cost target for dispensed hydrogen; it is instead considered within the scope of the U.S. DOE R&D program. The DOE target is based on a calculation of threshold costs which shows that the total cost of dispensed hydrogen needs to be less than \$4/GGE—same as a kg of H_2 on energy content basis (in 2007\$)—to be competitive with other transportation options expected in 2020 (Ruth and Joseck, 2011). As shown in Table 3-10, of the total cost of dispensed hydrogen, roughly half is associated with production, and the other half with delivery and dispensing (including fueling station cost).

The calculation makes certain assumptions with regard to the incremental cost of HFCVs based on projected improvements in vehicle technologies (both HFCVs and the competing HEVs), as well as the projected gasoline price in 2020 and fuel economy (mileage for HFCVs and HEVs). Since Ruth and Joseck's report was published in 2011, there have been changes in the base case values. For example, the calculation uses fuel economy estimates of 59 miles/GGE for HFCVs and 42 miles/gal for HEVs. However, as per the recent presentation to the committee by Toyota, the HEV (Prius) fuel economy has improved from 42 mpg in 2010 to approximately 55 mpg in 2015 (Wimmer and Gazelle, 2016). The threshold cost calculations will need to be revised based on expected future improvements in the fuel economy of HEVs.

To avoid the high cost of delivery, on-site reforming has been evaluated in multiple demonstration projects as an option. However, current low demand has precluded a favorable cost benefit that could be realized through high-volume manufacturing and standardization of components. Also, on-site hydrogen generation by reforming may face operational challenges, and there may be greater need to store excess hydrogen at the fueling station to manage supply/demand balance.

Electrolysis, seen as the ultimate pathway for renewable hydrogen production, is still an expensive option. The basic technology for the water electrolysis

TABLE 3-10 Apportioned Target Costs for Centralized and Distributed Hydrogen (H_2) Production and Delivery in 2020

	Centralized (\$/GGE)	Distributed (\$/GGE)
H_2 production cost[a]	1.90	2.30
H_2 delivery[b] cost[a]	2.10	1.70

[a] Based on a maximum hydrogen threshold cost of \$4.00/GGE.
[b] Dispensed, but untaxed.
NOTE: GGE, gasoline gallon equivalent.
SOURCE: Weil et al. (2012).

pathway is commercially well established. Current efforts to reduce the capital cost and improve the overall efficiency of electrolyzers to meet the production cost target are focused on improvements in the MEA, flow fields, gas diffusion layers, bipolar plate design, catalysts, membranes, etc. (Satyapal, April 2016; presentation to the committee, slide 21). The DOE Multiyear R&D plan (e.g., FCTO MYDD Production, 2015, Table 3.1.2) also points out materials development and improvements in manufacturing processes, as well as reduction in cost of electricity needs to meet the cost targets for renewable hydrogen production by electrolysis. This is challenging, since the scope for capital cost improvement is limited by performance and safety requirements and the efficiency improvement is limited by thermodynamic and other system constraints. Current electrolysis units are already highly efficient (at approximately 50 kWh/kg H_2 compared to a thermodynamic limit of 39.73 kWh/kg H_2), and any gains will be only incremental. As a consequence, even with the DOE long-term efficiency target of 43 kWh/kg H_2, the cost of electricity to run the unit is the most critical (greater than 50 percent) cost factor. Furthermore, electrolysis plants do not enjoy the same economy-of-scale benefit as is possible for chemical plants, such as for natural gas reforming or gasification of biomass and/or coal, due to its modular additive nature. This is evident in the capital cost projections for central and distributed electrolysis plants (Ainscough et al., 2014). As a result, as shown in Table 3-11, the projected future hydrogen costs for the base case electrolysis plants are about the same (\$4.23/kg H_2—forecourt; \$4.20/kg H_2—central), and are significantly higher than the 2020 target of \$2.30/kg H_2.

With increased use of renewable electricity, hydrogen production by electrolysis and its storage and use may be an economical pathway to manage the irregular availability of renewable sources like solar and wind. If so, the use of such "off-peak" electricity could make hydrogen produced by electrolysis cost competitive.

The cost target for the delivery portion of the total hydrogen cost is less than \$2/kg H_2 by 2020. The 2015 status, as per the hydrogen delivery technical team (HDTT) report, suggests that the lowest cost option at present is gaseous

TABLE 3-11 Hydrogen (H_2) Production High-Volume Cost Projections for Polymer Electrolyte Membrane Electrolysis Cases

Case Study	Low Value (\$/kg H_2)	Baseline (\$/kg H_2)	High Value (\$/kg H_2)	Early Market (\$/kg H_2)
Forecourt: Current case	4.79	5.14	5.49	5.79
Future case	4.08	4.23	4.37	—
Central: Current case	4.80	5.12	5.45	—
Future case	4.07	4.20	4.33	—

SOURCE: Ainscough et al. (2014).

hydrogen (GH2) tube trailers with projected cost of \$3/kg H_2 (Gupta and Soto, 2016). However, GH2 has payload limitations. So, as the demand grows, its utility is limited and other options need to be developed further to make them cost competitive. With the liquid hydrogen (LH2) pathway, the current liquid tanker delivery cost is projected to be \$3.80/kg H_2, but hydrogen liquefaction is the most energy consuming and consequently the most expensive step. It is therefore the focus of the DOE R&D goals and targets. Novel liquefaction technologies are being developed to reduce the cost of liquefaction.

However, with the high intrinsic energy demand for hydrogen liquefaction requiring very low temperature (20 K), it is a challenging task. Another approach being pursued under the DOE program, called cryo-compressed storage, uses the higher density of hydrogen at lower temperatures to store more hydrogen by compressing and keeping cold hydrogen in the vessel. If compressed cryogenic onboard hydrogen storage is successful and the chosen option, the same technical approach would apply to hydrogen delivery.

Currently, a major part of the delivery cost is the cost of the fueling station with 700-bar capacity, and of that, the compression cost constitutes a significant portion, as shown in Figure 3-13, which presents the components of cost. DOE R&D efforts for this near-term approach are therefore focused on reducing the cost of compression and other station costs, such as ground storage.

Based on the progress to date and the remaining gaps to achieve the ultimate cost targets, it is apparent that novel technologies are necessary to overcome the

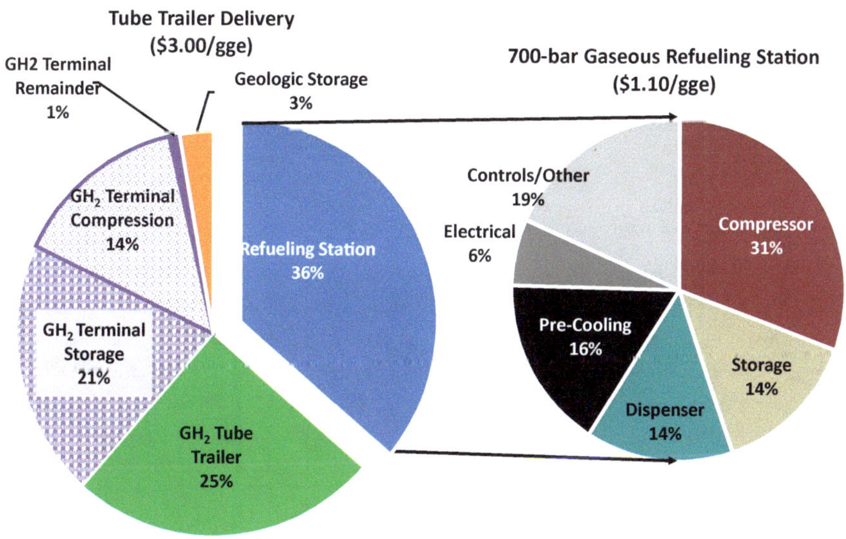

FIGURE 3-13 Components of cost for a hydrogen system.
SOURCE: Gupta and Soto (2016).

challenges to reduce the cost of hydrogen storage and delivery. One approach mentioned in the HDTT Roadmap relates to liquid hydrocarbon carriers (U.S. DRIVE, 2013f), although there is not much discussion of the pros and cons of such an approach. There were some early efforts within DOE, notably a concept developed by Air Products using carbazol compounds, which was later found to be impractical for technical reasons. However, this type of technology approach has not received sufficient attention within U.S. DRIVE, and a detailed techno-economic analysis along with in-depth exploration of the chemistry of this approach may be worth revisiting. According to a recent report, a German company called Hydrogenious Technologies (HT) has launched a commercial H_2 storage and logistics system using an innovative liquid organic hydrogen carrier technology (CryoGas International, 2016). Details of chemistry used, storage capacity, cost, and so on, are not provided. To enter the U.S. market, HT has signed an agreement with a U.S. hydrogen distribution company, United Hydrogen Group, for supply of these systems.

Another approach being investigated is to inject hydrogen in existing natural gas pipelines, which can then be withdrawn at desired locations after separating it from natural gas. This approach has its own challenges in terms of pressure management, pipeline integrity (materials compatibility), hydrogen separation and purification, public safety, and the associated costs (Melaina et al., 2013). This topic has not been addressed by the HDTT.

Progress and Key Achievements

U.S. DRIVE provided a list of DOE-funded hydrogen production and delivery projects relevant to the Partnership (see Table 3-12). The committee reviewed these projects presented at the 2015 and 2016 AMRs and has summarized the findings below along with a general discussion of the U.S. DRIVE technical team role and activities.

Hydrogen Production

The HPTT of the Partnership meets regularly and appears to have well-coordinated efforts to evaluate various production pathways and assess techno-economic viability. HPTT monitors a portfolio of technologies, as shown in Figure 3-14, which addresses near- and long-term needs. HPTT has identified appropriate technology areas that require focused R&D, and with their guidance along with other inputs, DOE is following a go/no-go methodology to downselect projects for continued funding. However, the committee feels that DOE could be more stringent about such decisions at various levels such as in technology areas and for specific projects. Examples of areas that may need more scrutiny are electrochemical compression, high-pressure electrolysis, and to some extent high-temperature electrolysis, as discussed later in this subsection.

TABLE 3-12 Department of Energy Projects on Hydrogen Related to U.S. DRIVE Goals

Project ID	Presentation Title	Organization
pd014	Hydrogen Delivery Infrastructure Analysis	Argonne National Laboratory
pd021	Development of High-Pressure Hydrogen Storage Tank for Storage and Gaseous Truck Delivery	Hexagon Lincoln
pd022	Fiber-Reinforced Composite Pipelines	Savannah River National Laboratory
pd025	Hydrogen Embrittlement of Structural Steels	Sandia National Laboratories
pd031	Renewable Electrolysis Integrated System Development and Testing	National Renewable Energy Laboratory
pd088	Vessel Design and Fabrication Technology for Stationary High-Pressure Hydrogen Storage	Oak Ridge National Laboratory
pd096	Electrolyzer Component Development for the Hybrid Sulfur Thermochemical Cycle	Savannah River National Laboratory
pd101	Cryogenically Flexible, Low-Permeability Hydrogen Delivery Hose	Nanosonic
pd102	Analysis of Advanced Hydrogen Production Pathways	Strategic Analysis, Inc.
pd103	High-Performance, Long-Lifetime Catalysts for Proton Exchange Membrane Electrolysis	Giner, Inc.
pd106	Reference Station Design	National Renewable Energy Laboratory
pd107	Hydrogen Fueling Station Pre-Cooling Analysis	Argonne National Laboratory
pd108	Hydrogen Compression Application of the Linear Motor Reciprocating Compressor	Southwest Research Institute
pd109	Steel Concrete Composite Vessel for 875 bar Stationary Hydrogen Storage	Oak Ridge National Laboratory
pd110	Low-Cost Hydrogen Storage at 875 bar Using Steel Liner and Steel Wire Wrap	WireTough Cylinders
pd111	Monolithic Piston-Type Reactor for Hydrogen Production through Rapid Swing of Reforming/Combustion Reactions	Pacific Northwest National Laboratory
pd112	Reformer-Electrolyzer-Purifier for Production of Hydrogen	FuelCell Energy, Inc.
pd113	High-Efficiency Solar Thermochemical Reactor for Hydrogen Production	Sandia National Laboratories
pd114	Flowing Particle Bed Solarthermal Redox Process to Split Water	University of Colorado

continued

TABLE 3-12 Continued

Project ID	Presentation Title	Organization
pd115	High-Efficiency Tandem Absorbers for Economical Solar Hydrogen Production	National Renewable Energy Laboratory
pd116	Wide Bandgap Chalcopyrite Photoelectrodes for Direct Solar Water Splitting	University of Hawaii
pd117	High-Temperature, High-Pressure Electrolysis	Giner, Inc.

SOURCE: Project numbers are from the Department of Energy's 2016 Annual Merit Review, see Cooper (2016a).

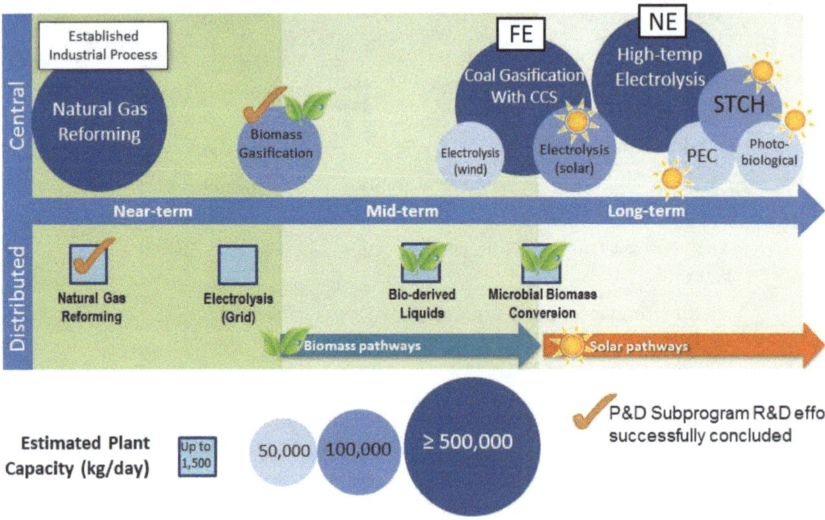

FIGURE 3-14 Technology pathway development timelines, feedstocks, and energy sources for hydrogen production.
NOTE: CCS, carbon capture and storage; P&D, production and delivery; PEC, photoelectrochemical; R&D, research and development; STCH, solar thermochemical hydrogen. FE and NE refer to R&D efforts in the Department of Energy's Offices of Fossil and Nuclear Energy, respectively.
SOURCE: Chapman and Randolph (2016).

According to the HPTT report, hydrogen production with natural gas reforming, centrally or distributed, can meet the production cost targets with current low natural gas prices (Chapman and Randolph, 2016). This pathway also results in significantly lower GHG emissions (approximately 50 percent) per mile driven compared to petroleum-based fuels. Further reduction in GHG emissions with this pathway would require CCSU since CO_2 is also produced in the natural gas reforming process, as mentioned before. Alternatively, pathways are needed to use with renewable resources, such as biomass conversion and electrochemical conversion using renewable electricity (e.g., wind, solar). Current costs of these pathways are much higher, and technology advances are necessary to bring the costs down. These are the conclusions of the extensive techno-economic analyses conducted by DOE (Chapman and Randolph, 2016). As a result, with the limited funding available, DOE program efforts are focused in these areas.

Based on the summary presentation at the committee meeting in February 2016, there has not been much progress reported in hydrogen production and delivery areas with respect to cost since the NRC Phase 4 review (Satyapal, 2016). Based on techno-economic analyses using the DOE H2A model, it was concluded that distributed natural gas reforming can meet the production cost target of less than \$2/kg H_2 with a natural gas (NG) price of up to \$5-6/GJ. To achieve lower GHG emissions with natural gas reforming it needs to be coupled with CCSU, which will add cost and has its own technical and economic challenges. Moreover, CCSU is only practical with large central plants; hence, distributed NG reforming is only a short-term solution. HPTT has evaluated the cost projection for CCSU (Chapman and Randolph, 2016). Accordingly, it constitutes a significant portion (\$0.79-\$0.96/kg H_2) of the total cost of hydrogen by reforming (\$1.58-2.25/kg H_2). Moreover, the CCSU cost estimates are yet to be validated as the technology is still not fully developed, and the long-term viability of CCSU needs to be confirmed.

Among renewable hydrogen pathway options, only central biomass gasification is projected to achieve production costs closer to but more than \$2/kg H_2 (the 2020 target). Other pathways, mainly electrolysis, continue to have significantly higher cost projections (\$4-5/kg H_2). A key achievement in this area within the DOE program is a 40 percent reduction in PEM electrolyzer capital cost compared to a 2011 baseline as a result of bipolar plate innovations (Chapman and Randolph, 2016). However, the analysis shows that this cost reduction has no significant impact on the total hydrogen production cost since the major factor in the overall cost is still the actual cost of electricity (OPEX) to split water, which is outside the scope of any electrolyzer developments. As a result, slide 7 of the same presentation (see Figure 3-15) shows that the estimated cost of H_2 production with PEM electrolysis has not changed much from 2010 to 2015 for distributed or central plants.

Recognizing this challenge, two approaches are being investigated—electrolysis at high pressure to save on compression costs and electrolysis at high temperature to save on electricity cost. High-pressure electrolysis is an ongoing effort, and various companies have in the past attempted to commercialize this approach. There is a limiting pressure beyond which the benefits of high pressure are offset by the higher capital costs and safety issues. Thus, high-pressure electrolysis efforts have shown limited success. Since the NRC Phase 4 report, no significant progress has been reported on this approach within the DOE program.

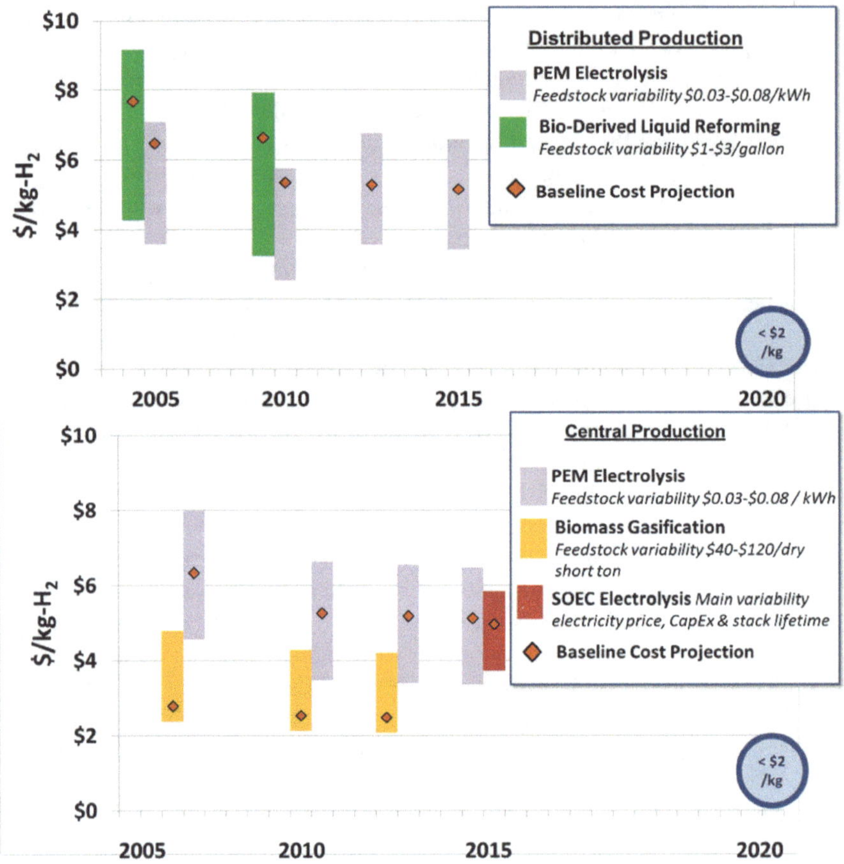

FIGURE 3-15 Estimates of cost of hydrogen production for different technologies.
SOURCE: Chapman and Randolph (2016).

An ongoing project by Giner, Inc. (AMR—pd117) has already concluded that high-pressure operation is generally uneconomical.

The other option proposed is high-temperature electrolysis (solid oxide electrochemical, or SOEC), wherein available, cheaper thermal energy is expected to reduce the electric energy requirements and hence the corresponding cost. The high temperature operation poses materials challenges and results in higher capital costs compared to PEM fuel cells. Hence, materials development is the focus of this program. The basic technology for this approach is the same as that for solid oxide fuel cells (SOFCs), which are being developed under the Solid State Energy Conversion Alliance program within the DOE Fossil Energy Office. Historically

SOFC developments date back more than 30 years, and a few years ago efforts to develop large-scale SOFC systems (multiple MW scale) were suspended by major developers due to insurmountable material stability issues.

James et al. (2015) of Strategic Analysis, Inc. (SA) provide cost projections for selected technologies, specifically PEM electrolysis, SOEC, and bio-fermentation, based on a detailed analysis. The results should help guide DOE decisions regarding focus areas and funding levels. Even with some aggressive assumptions with regard to potential reduction in capital cost and increase in efficiency for a large SOEC system (50 ton/day central plant), H_2 cost is projected to be near $4/kg, and it is shown that the key cost factor is the cost of electricity, making up more than 50 percent of the total cost, same as with PEM electrolysis. In view of the prior experience with such systems and the cost estimates, renewed efforts to conduct R&D of SOEC with federal funding are questionable and need to be reevaluated. Nevertheless a recent funding announcement from DOE shows that they are still funding three new projects (Ceramatec, FCE, and Giner) for the same general technology approach. One option to address the high cost of electricity required for hydrogen produced by electrolysis and to significantly reduce the cost of renewable hydrogen production is to use "excess" electricity produced by wind or solar, which is much cheaper; for example, it is estimated at $0.02/kWh or less compared to the average U.S. electricity cost of ~$0.11/kWh (see the section "Electricity as an Energy Source and the Grid," later in Chapter 3). This approach, generally known as power-to-gas, is being pursued more aggressively in Europe, especially in Germany. In the United States, the first such project started in mid-2015 with NREL teaming with Southern California Gas Co. and National Fuel Cell Research center. ARPA-E recently announced funding of $2 million to Dioxide Materials for a similar project using their proprietary catalyst-based electrolyzer technology. A greater emphasis on this area would help overcome the electricity cost barrier to electrolysis mentioned earlier. A recent initiative with DOE leadership, called H2@Scale, includes this approach, along with a cost analysis, and shows the potential to reduce the cost of hydrogen by >45 percent to approach the target of $2/kg (HTAC presentation, April 6, 2016; IPHE Meeting, May 20, 2016).

A similar analysis of a biofermentation plant has resulted in an estimated H_2 production cost of $4.62/kg, assuming by-product credits. However, revised analysis by SA presented at the 2016 AMR showed the projected H_2 costs to be $58.53/kg and $5.65/kg, respectively, for the current and future cases. The main difference between the two cases is the assumption that the corn stover concentration of 175 g/L can be achieved in the future compared to 5 g/L for the current case. In this case, while the major cost contributor for the current case is the cost of heat required, capital and feedstock costs are the major factors for the future case with higher efficiency and heat generation by burning lignin. Given the aggressive assumptions in the preceding analyses, it seems challenging to achieve even the projected costs, and perhaps impossible to meet the 2020 DOE cost target of $2/kg with any of these options.

Based on these analyses, it is evident that a novel technology is needed to meet the cost targets or the cost target needs to be revised to reflect realistically plausible cases. In either case, a paradigm shift is required. In 2015-2016, DOE funded development of two new technology concepts, and SA has undertaken analysis of these to estimate corresponding costs. The two technologies are (1) bio-oil reformation using a monolith catalyst in a dual bed system with swing operation; and (2) NG reformation in a molten carbonate-based electrolyzer operated at approximately 600°C temperature, with simultaneous CO_2 removal to directly obtain approximately 98 percent purity H_2.

The committee understands that these are high-risk technology projects. However, lessons learned from the past should not be overlooked. For example, in the case of bio-oil reforming, it should be noted that the basic concept of conducting reforming and regeneration in a cyclic mode was extensively investigated before (e.g., GE, ExxonMobil, BOC). Even with natural gas feed, there were operational issues with thermal management, stability, contamination, and so on. Similarly, the reformer-electrolyzer-purifier approach is a variation on past efforts to remove CO_2 in situ in a reformer, which have not been so successful. SA projections should help DOE to determine viability of these approaches in terms of potential to meet the cost target.

Other long-term production pathways include wind-based electrolysis, photoelectrochemical (PEC), solar thermochemical hydrogen (STCH), and photobiological (PB) approaches; these are in early stages of development, with low efficiencies and high costs, and face more daunting challenges in terms of achieving the cost targets. Efforts in these areas are primarily focused on improving conversion efficiency, which is currently the main barrier. If target efficiencies are achieved, the pathways can be attractive since all of these use renewable energy and therefore lead to very low GHG emissions. Pathway analysis conducted by NREL shows that the projected cost of hydrogen production using the PB approach is very high (more than \$9/kg) compared to other approaches mentioned (e.g., STCH: less than \$4/kg) (Ramsden, 2015). Novel reactor and material concepts being developed for STCH at Sandia National Laboratories (McDaniel and Ermanoski, 2015) and by Professor Weimer at the University of Colorado Boulder (Weimer et al., 2015) show promising results. The approach is based on a two-step process wherein metal oxide is first heated by concentrated sunlight to high temperatures to reduce it and produce O_2. In the second step, the reduced material is exposed to H_2O to reoxidize it to its original oxidation state and produce H_2.

The materials and process conditions are being optimized in these projects to develop a more reliable and cost-effective process. Weimer's approach consists of operating the reactor in a circulating mode similar to a fluid catalytic cracking reactor. This is challenging in terms of optimum reaction kinetics (material selection), material stability, and thermal management, but if successful, it has many benefits in terms of scale-up and economics.

If the final analysis confirms the preliminary results and if the scope for further improvement is limited with some of the approaches like PB, it would make sense to stop or scale back efforts on those that appear less promising and focus more on promising approaches like STCH. Another attractive approach is the microbial biomass conversion as it presents the opportunity to purify water simultaneously. This approach has not been included in the NREL study, but is being developed at NREL in collaboration with Penn State (Maness and Logan, 2015). There is also a new project begun in 2016 at Oregon State University (PD129) to develop a similar technology. These projects are not included in the list of U.S. DRIVE relevant projects. These are still early stage developments and should be pursued to advance the technology, albeit with an eye on economic viability. These findings are consistent with those of the HPTT.

Hydrogen Delivery

DOE-funded R&D projects are focused on improvements in high-pressure component performance and materials for low-cost fabrication. Efforts to reduce cost are largely through developments of new and improved methods for producing carbon fibers, which are used for the fabrication of lightweight, high-strength vessels for high-pressure hydrogen storage and transport as well as pipelines for hydrogen transport as a long-term option. Since high-pressure hydrogen transport has limitations in terms of pressure and payload capacity, there are also efforts to develop alternative approaches such as LH2 delivery. More recent efforts in hydrogen storage are focused on cold compressed hydrogen, called cryo-compressed, and use of adsorbents, especially in combination with low temperatures. Significant progress has been made in developing metal-organic framework (MOF) materials which, when used at cryogenic temperatures, can store high amounts of hydrogen. More discussion on hydrogen storage technologies can be found in the section on onboard hydrogen storage of this report. The same technologies are likely applicable for cost-effective hydrogen transport and delivery.

As mentioned earlier, the current cost of hydrogen at the pump is high ($13-16/GGE), and while high-volume usage will bring the costs down, to meet the ultimate cost targets advancements are required in technologies for hydrogen storage, delivery, and dispensing. Even with future projections, the cost of hydrogen delivery is a major contributor to the total cost of dispensed hydrogen. The delivery includes hydrogen processing (e.g., compression) and handling at the production facility, its transport to the fueling station and fueling station itself, which encompasses storage, further compression, and metering the hydrogen into the vehicle tank. All of these steps operating in tandem present operational complexities and add significant costs. DOE has conducted an analysis of key cost factors to identify and focus R&D efforts on critical areas. While the overall budget for the delivery area may be limited, the committee feels that the funds are being used appropriately with prioritization of R&D topics within the current delivery approaches, and steady progress is being made. However, to meet the ultimate cost targets, a radically different approach may be necessary to overcome challenges and limitations of the high-pressure storage and transport pathway.

Within the current framework, a key achievement since the NRC Phase 4 report relates to a 25 percent reduction in the cost of gaseous hydrogen delivery from approximately \$8 to approximately \$6/kg H_2 with the development of high-pressure (500-bar) tube trailers by Hexagon to replace the conventional ones with less than 200 bar pressure, thereby increasing the payload, which results in the reduction of cost per kg H_2 (U.S. DRIVE, 2013f). Another accomplishment in the delivery arena is the demonstration by Argonne National Laboratory of a tube trailer consolidation strategy to lower forecourt compressor capital cost by about 60 percent (Gupta and Soto, 2016). Also Oak Ridge National Laboratory (ORNL) is developing a steel/concrete composite vessel for hydrogen storage on site at distribution terminals as well as at fueling stations, which can help reduce the overall delivery cost.

With the LH2 delivery pathway, seen as a near- to mid-term option, although delivery of LH2 is the most cost effective, liquefaction is the most energy consuming and expensive step. No significant progress has been reported since the NRC Phase 4 review for this pathway. But two new projects have been funded to develop novel concepts. One at NREL (Ainscough et al., 2014) involves use of a vortex tube for separation and simultaneous refrigeration to increase the efficiency of ortho/para hydrogen conversion, which is the most energy consuming step in hydrogen liquefaction. The other project at PNNL (Holladay, 2016) involves magnetocaloric refrigeration to improve the efficiency of liquefaction. This concept was originally developed by Prometheus with DOE funding. It remains to be seen whether the improvements envisioned in the new projects lead to a successful technology with potential to significantly reduce the cost of liquefaction.

A DOE-funded project (2007-2015) led by Fuel Cell Energy (FCE) attempted to develop electrochemical hydrogen compression (EHC) to reduce compression cost. While they improved cell durability, capacity, and efficiency, the final pressure achieved under practical conditions was limited to 3,000 psi (approximately 200 bar). Also, with an electrochemical cell as the building block, when scaled up, the compressor does not have the same economy-of-scale benefit as the mechanical compressors. Thus, the compressor cost can be a challenge. In the NRC Phase 4 report, it was suggested that the Partnership should evaluate this approach and compare it with conventional compression (NRC, 2013, p. 129). However, this has not been mentioned anywhere in the recent Partnership reports and presentations. Yet, surprisingly, there is a recent new DOE funding award announced to Giner, Inc., for developing EHC. Another project awarded to Greenway Energy is based on a novel approach combining two technologies: EHC and metal hydride compression (MHC). Development of MHC was attempted before by Ergenics and others; lessons learned from those projects should be helpful in developing the new concepts.

Independently, without any DOE funding, Linde has developed a novel hydrogen compression technology using ionic liquids to replace the mechanical piston, which has wear and tear issues (Mayer, 2014). The ionic liquid can be pumped to high pressures with much less energy, which results in lower energy costs. Also with the use of an immiscible liquid, the liquid can be in direct contact with hydrogen and the compression can be nearly isothermal and without any contamination issues. As a result, the compressor has high efficiency and throughput, and it can compress hydrogen from ambient to 900 bar with a single five-stage compressor. The current model IC90, for example, has a maximum delivery rate of 33.6 kg/hr H_2 and maximum pressure capacity of 1,000 bar. With input H_2 at 5 bar, power consumption is 75 kW, with specific energy consumption of only 2.7 kWh/kg H_2 (a 40 percent energy saving compared to a conventional dry piston compressor). With mass production, the current capital cost is expected to be reduced by almost 50 percent. Initially the technology was used in compressed natural gas fueling stations, and then extended to hydrogen compression starting around 2010. Linde has now commercialized this technology, and it is used in their standard hydrogen fueling station design (Mayer, 2014). Linde also recently started a serial production line for these products. The DOE is well aware of and familiar with this technology and the potential benefits. It would be useful for the Partnership to review and compare this compression technology with the ongoing DOE-funded programs to calibrate future R&D needs and direction.

Current DOE R&D efforts aimed at reducing fueling station costs include development of a linear motor reciprocating compressor to lower maintenance costs by minimizing wear parts and of coatings for compressor seals to improve durability in high-temperature, high-pressure hydrogen. R&D efforts are also focused on improving hydrogen dispensing equipment at the station, such as durability of the dispensing hose at cryogenic temperatures. The projects highlight challenges related to materials compatibility, compression efficiency, durability, maintenance, and associated costs. Some new approaches are emerging, such as underground or canopy storage, and high-pressure delivery and storage to eliminate compression at the fueling station, thereby reducing station footprint and cost. As pointed out before, there is clearly a need to strive for novel approaches to make hydrogen an affordable and practical fuel of choice.

In the DOE program, the long-term delivery cost projections are based on medium-pressure hydrogen gas delivery from a large central plant via pipelines. Various efforts are under way to reduce the cost of pipeline delivery (Gupta and Soto, 2016). These include development of high-volume centrifugal compressors and fiber-reinforced composite pipelines to reduce the distribution cost as well as safety-related studies to understand hydrogen embrittlement of structural steels. Hydrogen embrittlement is well known in the industry and is well studied. It may be worth making sure that there is no duplication of efforts, so that funding can be used for other more pressing matters.

Significant Barriers and Issues That Need to Be Addressed

The key remaining barriers and issues identified for the hydrogen production and delivery area discussed above are summarized here:

- A critical barrier to meeting the hydrogen production cost target with electrolysis is the cost of electricity to run the unit, which cannot be resolved with incremental improvements in capital cost and efficiency targeted with the current R&D.
- In the long term there is need to produce hydrogen with no or very low GHG emissions. This requires use of renewable sources such as biomass or solar energy, directly or indirectly through electrolysis of water. Efficiencies of various renewable pathways are currently low. They need to be significantly higher for cost-competitive hydrogen production.
- While high-pressure (350 and 700 bar) storage is currently the only viable option, it is a high-cost option when combined with the required delivery and dispensing infrastructure from a gaseous or liquid hydrogen source. The high cost of materials such as carbon fibers, required for high-pressure storage, and the high cost of compression continue to be the significant barriers that need to be addressed.
- Pipeline delivery of high-pressure hydrogen is envisioned as the long-term approach. This entails high initial investment costs as a key barrier along with land access and public safety as well material compatibility, durability, and costs.
- One critical issue with hydrogen as a light, colorless, odorless gas is that it can easily leak without easy detection from piping, fittings, valves, equipment, etc. especially under the high-pressure conditions. This can result in loss as well as unsafe conditions due to the flammable and explosive nature of hydrogen. This leads to the use of more expensive equipment.
- An obvious key barrier for the hydrogen production and delivery is the inability of the current technology options to meet the cost targets to be competitive with other transportation options.
- An overarching barrier is inadequate funding for the hydrogen production and delivery program to accomplish the stated goals within the desired time frame to make an impact. As a consequence, there is insufficient focus on novel generation technologies to overcome the technical barriers.

Responses to Recommendations from the NRC Phase 4 Review

NRC Phase 4 Recommendation 4-3. While a hydrogen from coal demonstration plant could address many of the downstream integration issues and thus provide more certainty around the probable capital costs, the Committee recommends that any hydrogen from coal demonstration should be paced (1) to match the pace and progress of commercial scale carbon sequestration, and (2) to support a mature hydrogen fuel cell vehicle fleet in the event that natural gas becomes too costly or unavailable.

Partnership Response. To clarify, the U.S. Department of Energy (DOE) supports research on Carbon Capture and Storage as a part of its portfolio of technology options to address global climate change. DOE's Office of Fossil Energy supports research on the conversion of coal to hydrogen. The current focus is on advanced hydrogen membranes and the use of hydrogen in gas turbines. Office goals include capturing carbon dioxide in the production of electricity, hydrogen, and other marketable products while meeting all environmental standards. The current portfolio consists of approaches that might not require any storage technologies.

Committee Assessment of Response to 4-3. The Partnership states that DOE has funded efforts to develop CCS in conjunction with NG reforming (SMR). The cost estimates for this option show that CCS adds substantial costs, even at a large-scale production plant. Further efforts in CCS development may bring these costs down. However, viability of CCS remains uncertain.

NRC Phase 4 Recommendation 4-4. Support should continue at the fundamental component level (e.g., catalysts, anode supports) for all types of electrolyzers as well as for associated power electronics.

Partnership Response. There has been significant progress in the development of proton exchange membrane (PEM) electrolyzer technology over the last decade including the demonstration of a more than 80% reduction in electrolyzer stack cost. As a result, the U.S. Department of Energy's (DOE) Fuel Cell Technologies Office (FCTO) is moving more of its electrolysis activity to the Technology Validation subprogram and choosing to be more selective with conventional electrolysis R&D. To avoid duplication of effort, FCTO continues to leverage fundamental electrolysis component R&D funded through Basic Energy Science, ARPA-E (Advanced Research Projects Agency-Energy), and Small Business Innovation Research (SBIR) awards, which are well suited to incentivize further innovations. Active awards focus on improving stack efficiency and lowering the cost of both alkaline and PEM electrolysis. FCTO also continues to leverage fundamental R&D of advanced power electronics in cross-cutting DOE initiatives for renewable energy applications (including "grid integration" and "wide bandgap semiconductor" initiatives). This research is applicable to high-efficiency electrolysis.

Committee Assessment of Response to 4-4. The Partnership has pointed out relevant projects and significant progress made toward reducing the capital cost of an electrolyzer. However, as discussed in this section, this cost reduction has not resulted in any significant reduction in the overall cost of hydrogen produced by electrolysis because of the fact that the major cost contributor is the cost of electricity.

NRC Phase 4 Recommendation 4-5. Technical development and systems analysis on high-pressure electrolytic hydrogen production should be supported to determine the costs, scalability, benefits, and developmental steps required to make it viable compared with conventional compression. With the goal of eliminating mechanical compression, additional work should be done on high pressure electrolysis that can produce pressures of 84 MPa to 98 MPa (12,000 to 14,000 psi) and have sufficient capacity to do a fast tank fill (3 minutes).

Partnership Response. Two U.S. Department of Energy (DOE) DOE Small Business Innovation Research projects addressing electrolytic home refueling (Giner, Inc., and Proton OnSite) have demonstrated the potential feasibility of high-pressure electrolytic hydrogen production at 5,000 psig. DOE, in coordination with its U.S. DRIVE partners, will continue system-level analysis to define the tradeoffs between increased stack production pressure and conventional compression and to help identify the optimum electrolyzer pressure. Within DOE's Fuel Cell Technologies Office, the Hydrogen Delivery subprogram is also funding developments in the closely related technology of electrochemical compression. Current projects include a Fuel Cell Energy electrochemical compressor, which recently demonstrated delivery of >12,500 psig hydrogen, and an analysis project with Strategic Analysis, Inc., in conjunction with Pacific Northwest National Laboratory, to create an electrochemical compression component model for the Hydrogen Delivery Scenario Analysis Model (HDSAM).

Committee Assessment of Response to 4-5. In the response, the Partnership pointed out ongoing DOE-funded projects conducted by Giner and Proton Onsite. It also mentions a Fuel Cell Energy project on electrochemical compression. Since the last report it is not clear how much progress has been made in this area, and the techno-economic viability of electrochemical compression seems uncertain. Yet, DOE has recently awarded funding to Giner for a new project using such a technology. The ongoing analysis by Strategic Analysis, Inc., should help make the determination. The committee feels that efforts should focus on exploring other novel approaches to compression. The recently funded project to Greenway combining EHC and MHC may be a step in the right direction.

NRC Phase 4 Recommendation 4-6. The U.S. DRIVE Partnership should continue to support the development, testing, and analysis of (distributed) renewable electricity production methods in combination with the electrolysis of water.

Partnership Response. The Partnership agrees that the development, testing, and analysis of renewable electricity production methods in combination with water electrolysis is important. In support of this view, the U.S. Department of Energy has addressed the integration of electrolysis with renewables through the Fuel Cell Technologies Technology Validation subprogram. One example of such support is the National Renewable Energy Laboratory's (NREL) Wind-to-Hydrogen project, in which wind energy is directly converted to hydrogen via electrolysis. NREL's wind site includes facilities for testing electrolyzers directly connected to power output from wind turbines and photovoltaic arrays in order to better understand issues specific to the operation of electrolyzers on variable power sources. In addition, NREL has been carrying out a wind-to-hydrogen cost modeling analysis effort focused on optimizing wind-based water electrolysis production, including analyzing 42 potential sites in 11 states. In addition, through its market transformation activities, DOE is partnering with the U.S. Department of Defense and the State of Hawaii to investigate the integration of electrolyzers with renewables, electrolyzer performance with intermittent power, and electrolyzer contribution to grid stability.

Committee Assessment of Response to 4-6. The suggested approach, generally known as power-to-gas, is being pursued more aggressively in Europe, especially in Germany. In the United States, the first such project started in mid-2015, with NREL teaming with Southern California Gas Co. and National Fuel Cell Research Center. ARPA-E recently announced funding of $2 million to Dioxide Materials for a similar project using their proprietary catalyst-based electrolyzer technology. A greater emphasis on this area would help overcome the electricity cost barrier to electrolysis mentioned above.

Findings and Recommendations

Hydrogen production by natural gas reforming is currently a cost-effective option for the near-term hydrogen requirements, and it also provides a pathway to reduced GHG emissions. To further reduce GHG emissions, the use of renewable sources of energy, such as biomass, wind, and solar, is required. Development of such technologies is the focus of the long-term R&D. However, delivery and dispensing of hydrogen is still prohibitively expensive and requires technological advances to meet the overall cost targets for the HFCV option to be viable in the future. Currently the accepted option for onboard storage is 700-bar compressed hydrogen. The delivery and dispensing of hydrogen needs to meet the corresponding requirements—that is, even higher pressure (e.g., 875-900 bar) at the pump. Thus, the R&D focus has been to develop low-cost compression technologies and materials and concepts for high-pressure hydrogen storage and transport. There are several hurdles with this approach, as pointed out, and alternative new concepts need to be continually developed.

Finding 3-22. U.S. DRIVE does not have a cost target for dispensed hydrogen; it is instead considered within the scope of the U.S. DOE R&D program. The DOE cost target for dispensed hydrogen of less than $4/kg H_2 is based on their calculation of threshold cost.[12] Since DOE calculated this cost in 2011, there have been changes in the base case values such as the fuel economy for hybrid electric vehicles.

Recommendation 3-10. The hydrogen threshold cost calculation, published by the Department of Energy in 2011, should be revised by taking into consideration the advances in competing hybrid vehicle technologies as well as any progress made with vehicular hydrogen fuel cells. This should be carefully assessed and addressed by the appropriate U.S. DRIVE teams as well as the Executive Steering Group to incorporate the implications in the Partnership plans.

Finding 3-23. Hydrogen production by natural gas reforming at large central plants can meet the DOE production cost target. However, to meet the greenhouse gas reduction requirement, it needs to be combined with carbon capture

[12] Threshold cost is calculated so as to be competitive with other transportation options that are expected in 2020 (Ruth and Joseck, 2011).

and sequestration, which adds significant cost and particularly entails uncertainty with regard to the long-term viability of CO_2 sequestration.

Finding 3-24. Even with aggressive assumptions with regard to potential reduction in capital cost and the increase in efficiency of an electrolyzer (including high-pressure and high-temperature versions), hydrogen production cost with large scale (50 ton/day H_2) or small-scale on-site plants is projected to be near \$4/kg H_2 at best, and it is shown that the key cost factor is the cost of electricity, making up more than 50 percent of the total cost, just as the situation is with PEM electrolysis.

Recommendation 3-11. U.S. DRIVE in conjunction with the Department of Energy should assess if the focus and prioritization of the electrolyzer research and development is appropriate considering the fact that the main cost factor is the cost of electricity and incremental improvements in capital cost and efficiency may not have as much impact on the cost of hydrogen produced.

Finding 3-25. Liquid hydrocarbon carrier technology for hydrogen delivery has not received sufficient attention within the U.S. DRIVE delivery technical team.

Recommendation 3-12. U.S. DRIVE should conduct a detailed analysis of the liquid hydrocarbon carrier option for hydrogen storage and transport, including potential new chemistries, and compare it with other pathways to determine if it merits development.

Finding 3-26. With the current pathways, hydrogen compression is a major cost item, which is being investigated. In addition to improvements in mechanical compression technologies, DOE has funded electrochemical compression projects. However, it has not been evaluated by the Partnership as suggested in the NRC Phase 4 report. Also, other competing options, such as ionic liquid based compression, have not been critically assessed by the Partnership to identify the scope for improvements and future R&D needs to meet the DOE cost targets.

Recommendation 3-13. As suggested in the National Research Council Phase 4 review, electrochemical compression should be evaluated and compared with other existing and emerging options (e.g., Linde's ionic compressor) to decide on the appropriate future development direction for compression R&D.

Finding 3-27. Although considerable efforts have already been expended to reduce the costs of hydrogen production, delivery, and dispensing with current pathways, it is clear that there are significant hurdles to meet the current cost target for dispensed hydrogen. Therefore, there is a need to continually seek novel technology approaches and innovative pathways.

Recommendation 3-14. In view of the relative costs of hydrogen production and delivery, research and development (R&D) progress to date and the remaining barriers discussed, in identifying R&D needs, the Partnership should address the need for increased emphasis on novel and innovative approaches to hydrogen delivery and dispensing technologies to reduce the overall cost of hydrogen.

HYDROGEN TRANSITION ISSUES

While the electricity infrastructure already exists, and competitive electricity costs and recharging station deployment are helping to bring BEVs and PHEVs to market more readily, the hydrogen refueling infrastructure needed for deploying a significant number of HFCVs is nonexistent and needs to be developed. There are existing technologies and networks for transporting hydrogen to customers in the industrial segment, and it is a starting point for vehicle fueling infrastructure. In addition to the large quantities of hydrogen produced today by and for refineries and for production of chemicals (ammonia, methanol), relatively large liquid hydrogen plants also exist to supply the industrial hydrogen markets (metals, glass, food processing, and other). These existing plants often have sufficient excess hydrogen capacity available, which can be used for hydrogen supply to fueling stations in the near term. However, as pointed out before, current costs of delivery (including fueling station cost) are generally high.

Today there are three options available to supply hydrogen to fueling stations: (1) transmission and distribution from centralized plants of gaseous hydrogen by tube trailers, (2) distribution of liquefied hydrogen by tankers, and (3) on-site generation of hydrogen at the fueling site using natural gas reforming or electrolysis of water (see Figure 3-16). However, current technologies have limitations, such as tube trailer pressure of no more than 200 bar, leading to maximum payload of only approximately 300 kg H_2 per trailer; liquid hydrogen boil-off losses; safety, codes, and standards issues as well as operational and maintenance issues for on-site natural gas reformers; and especially high cost of delivery combined with station costs for compression and storage, relative to the current DOE cost target for dispensed hydrogen at fueling stations.

The fuel pathway integration technical team (FPITT) has representatives from DOE, four energy companies, and the NREL (Joseck and Verduzco, 2016). Since the Phase 4 NRC review carried out during 2012, the team now includes Ford as a member and Air Products and Chemicals, Inc. as an associate member. As noted by Joseck and Verduzco (2016) and in the FPITT roadmap (U.S. DRIVE, 2013d) this technical team supports the U.S. DRIVE Partnership in the identification and evaluation of implementation scenarios for fuel cell technology pathways, including hydrogen and fuel cell electric vehicles in the transportation sector, both during a transition period and in the long term. As stated, the FPITT charter is to review publicly available, ongoing and completed analyses of hydrogen and fuel cell technology pathways, to provide industry-based perspective on R&D

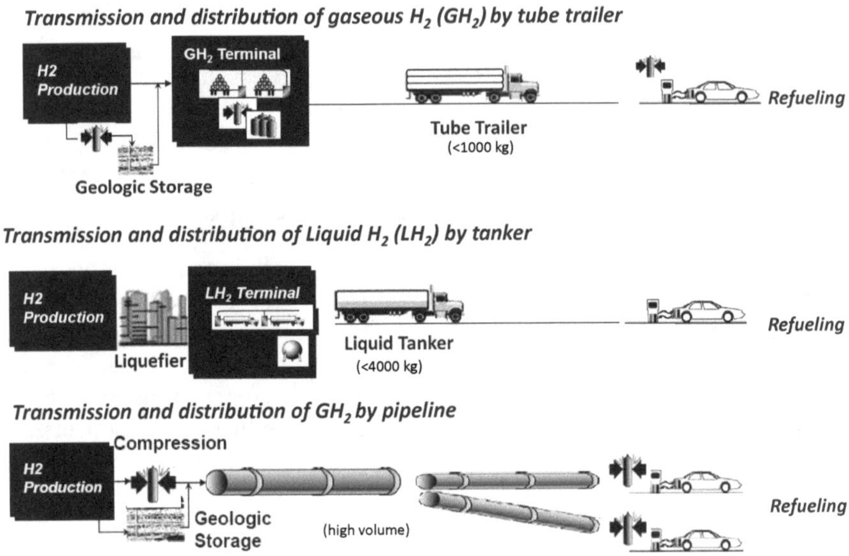

FIGURE 3-16 Options for hydrogen supply to fueling stations.
SOURCE: Gupta and Soto (2016).

needs, targets, direction, and potential ramp-down for consideration by DOE in the management of the DOE FCTO. The FPITT is aligned with the Partnership Goal 2: "Enable reliable fuel cell electric vehicles with performance, safety, and costs comparable to or better than advanced conventional vehicle technologies, supported by viable hydrogen storage and the widespread availability of hydrogen fuel" (U.S. DRIVE, 2016).

In view of the preceding, the FPITT goals are as follows (U.S. DRIVE, 2013d):

- Assess techno-economic and environmental benefits of integrated hydrogen production, delivery, dispensing, and use pathways and compare them to other fuel pathways;
- Identify and bridge knowledge gaps that limit the ability to evaluate implementation scenarios for fuel cell technology pathways;
- Promote transparency in analysis activities by providing feedback on the communication of analysis parameters and results;
- Identify technical and institutional barriers to implementation of the pathways; and
- Promote consistency between goals, analysis, and modeling efforts of various technical teams of the Partnership to ensure consistent overall targets and effective use of resources.

As noted in the FPITT roadmap (U.S. DRIVE, 2013d), hydrogen transportation fuel pathways will need to compete against an existing petroleum-based fuel infrastructure that is well understood, highly efficient, and has widespread consumer acceptance. With the continual introduction of new vehicle technologies, the market conditions in which hydrogen must compete are continually evolving. Therefore, one of the most important technical challenges in implementing hydrogen as a transportation fuel is estimating the costs, benefits, and risks of potential hydrogen fuel pathways. The FPITT has done an excellent job of identifying gaps and barriers, which are listed as follows:

- Early transition station cost analysis;
- Well to wheels and life-cycle assessment: (1) energy, petroleum use, and greenhouse gas emissions for hydrogen pathways with 10,000 psi (700 bar) onboard storage; and (2) more complete life-cycle assessment of energy and emissions for hydrogen pathways;
- Marginal abatement cost of carbon;
- Hydrogen dispensing pressure optimization analysis; and
- Technology status/TRL and early to mature H_2 markets scenario analysis.

DOE has provided funding to undertake tasks to address some of these gaps and barriers. Progress has been made in many of the areas. A methodology was developed to evaluate complete hydrogen production, delivery, and dispensing pathways and to report on life-cycle hydrogen cost, energy use, and greenhouse gas emissions (Chapman and Randolph, 2016). NREL has completed evaluation of several current and future pathways, and the results have been published. FPITT along with ANL and ORNL evaluated trade-offs between hydrogen dispensing pressure with associated station costs and emissions and customer convenience. The study has not been conclusive and requires further analysis.

The Partnership provided a list of DOE-funded projects as those relevant to U.S. DRIVE. However, many of the projects listed were completed in 2015 with inconclusive results (see Table 3-13). Models and tools are well developed, but analysis requires reliable data, which are lacking.

The program overview states that while the 2020 at-the-pump H_2 cost target is still less than \$4/kg, a new early market cost target of \$5-10/kg has been introduced (Joseck, 2016). The cost analysis on a life-cycle basis shows that many of the options, except for BEVs, have levelized driving costs in the range \$0.35-0.40/mi (BEV: \$0.50-0.55/mi). Also, on a \$/mi basis, H_2 fueling infrastructure costs are close to but higher than those for BEV charging. It is projected that the HFCVs can be cost competitive with conventional ICE vehicles on \$/mi basis by 2025-2030 time frame, if DOE cost targets for hydrogen and fuel cells are met. However, as mentioned in the previous section, the comparison basis used for H_2 production and delivery threshold cost is the HEV. There is a need to have a common basis to compare different results. Nevertheless, the combined envi-

TABLE 3-13 Department of Energy Projects Related to Hydrogen Pathway Issues Relevant to U.S. DRIVE Goals

Project ID	Presentation Title	Organization
sa033	Analysis of Optimal Onboard Storage Pressure for Hydrogen Fuel Cell Vehicles	Oak Ridge National Laboratory
sa035	Employment Impacts of Infrastructure Development for Hydrogen and Fuel Cell Technologies	Argonne National Laboratory
sa036	Pathway Analysis: Projected Cost, Life Cycle Energy Use, and Emissions of Emerging Hydrogen Technologies	National Renewable Energy Laboratory
sa039	Life Cycle Analysis of Water Consumption for Hydrogen Production	Argonne National Laboratory
sa044	Impact of Fuel Cell System Peak Efficiency on Fuel Consumption and Cost	Argonne National Laboratory
sa045	Analysis of Incremental Fueling Pressure Cost	Argonne National Laboratory
sa047	Tri-Generation Fuel Cell Technologies for Location-Specific Applications	University of California, Irvine
sa050	Government Performance and Results Act Analysis: Impact of Program Targets on Vehicle	Oak Ridge National Laboratory
sa051	Infrastructure Investment and Finance Scenario Analysis	National Renewable Energy Laboratory
sa052	The Business Case for Hydrogen-Powered Passenger Cars: Competition and Solving the Infrastructure Puzzle	University of Chicago
sa053	Retail Marketing Analysis: Hydrogen Refueling Stations	Kalibrate
sa054	Performance and Cost Analysis for a 300 kW Tri-Generation Molten Carbonate Fuel Cell System	Argonne National Laboratory
sa055	Hydrogen Analysis with the Sandia ParaChoice Model	Sandia National Laboratories
sa056	Status and Prospects of the North American Non-Automotive Fuel Cell Industry: 2014 Update	University of Tennessee

SOURCE: Project numbers are from the Department of Energy's 2016 Annual Merit Review, see Cooper (2016a).

ronmental and life-cycle analyses demonstrate the need for a portfolio approach. With respect to hydrogen, fuel availability is a major concern for early adopters. Therefore it is evident that while more efforts are required to bring down hydrogen fueling costs, funding and innovative financing efforts are particularly needed to promote hydrogen infrastructure during the transition phase. Also there is a need to identify and promote specific situations and scenarios with viable business cases to enable gradual growth of the infrastructure.

One example of some success with the use of fuel cells in vehicles is that of forklift trucks, primarily due to productivity gains and battery versus fuel cell life and maintenance costs. The financial benefits (e.g., labor and time savings) of using fuel cells instead of batteries are such that it seems to support the current high cost of dispensed hydrogen, albeit in a limited number of cases. BMW, for example, now operates more than 350 forklifts at their production facility in Spartanburg, South Carolina, using hydrogen fuel cells. Fuel cell ownership cost appears to be still an issue that is prohibiting widespread use of hydrogen and fuel cells in this application.

For passenger vehicles, which are the focus area of the U.S. DRIVE Partnership, commercialization is even more challenging. First, the vehicles are still quite expensive and have only just started to appear in the market with limited availability. Even more important is the lack of availability of hydrogen fuel at affordable cost for public use. Nevertheless, some progress is being made in establishing a hydrogen infrastructure in limited areas. A notable example is that of funding by the state of California for fueling stations. Similar efforts are now under way in the Northeast states, such as Connecticut and New York. But in both cases there is little, if any, involvement of the Partnership members. Toyota and Honda are supporting the infrastructure build-out by providing funding. For example, Toyota ($7.6 million) and Honda ($13.8 million) announced funding to First Element Fuel for installing fueling stations in California. Air Liquide (AL) recently announced development of a network of fueling stations in the Northeast corridor using LH2 produced in their plant in Quebec using mostly hydroelectric (fossil-free) power (Edwards, 2016). Under the special circumstances, including available low-cost LH2 in the region with greater than 100 mile transportation, use of existing facilities, and collaboration with Toyota for certain guaranteed hydrogen demand, AL has been able to have a viable business case. Efforts like these are helpful in transitioning to a growing market for HFCVs and should be sought for and encouraged. Nevertheless, widespread use of HFCVs will require availability of cost-competitive hydrogen anywhere in the country where vehicles are to be sold.

Complete fueling station cost analysis was conducted using factors developed for economies of scale and for the learning curve (Joseck and Verduzco, 2016). Based on the current single 100 kg H_2/day station cost, and using these factors reflecting capital cost and O&M cost savings, station cost for a scenario with 100 stations at 1,000 kg H_2/day each, is estimated to be just under $4/kg, which is still significantly higher than the target of $0.70/kg. This gap needs to be addressed with R&D which is a major challenge. The FPITT has also initiated an analysis, in conjunction with the C2G working group, to assess the marginal cost of carbon abatement for the different pathways. Similar other activities to address the gaps identified, such as 2 and 5 listed above, are in progress. Results of these analyses are critical to provide feedback to DOE for funding targeted R&D, to understand the viability of hydrogen pathways, and to assess commercialization timeline.

Japanese and Korean OEMs have begun introducing vehicles in the U.S. market, but lack of adequate hydrogen infrastructure represents a major roadblock in the market introduction and proliferation of HFCVs. Goal 2 in the U.S. DRIVE plan (U.S. DRIVE, 2014) quoted above has not received the attention it merits as the core industrial partners (e.g., the energy companies) do not seem to be sufficiently engaged in activities related to hydrogen production and delivery pathways. For this critical requirement, the U.S. DRIVE program is almost entirely dependent on the DOE and national laboratory efforts. However, the hurdles identified above suggest that apart from economies of scale and learning by doing, significant technological advances are necessary to meet the cost target for the dispensed hydrogen. This suggests that the current level of funding for the hydrogen program (production, delivery, station development) may not be adequate to meet the challenge. In particular, with regard to infrastructure development, the DOE budget of $2-3 million per year is focused on analysis. While this is an essential activity, it is not sufficient to promote the necessary infrastructure build-out activities. While some electric utility companies, representing electricity infrastructure, are part of the core U.S. DRIVE team, the same is not true for hydrogen; industrial gas companies, with hydrogen infrastructure expertise, are represented in the U.S. DRIVE Partnership only as one-off associate members of the various technical teams. DOE has initiated another partnership called H2USA, which has a wider membership and is almost entirely focused on hydrogen infrastructure. These efforts are complementary and helpful in promoting the U.S. DRIVE agenda.

Significant Barriers and Issues That Need to Be Addressed

A critical barrier to the widespread deployment of HFCVs is the lack of a hydrogen infrastructure. Although a few select states have been funding and promoting installation of public fueling stations, there are very few stations operating today. While companies like Toyota and Honda are providing additional funding for fueling stations, similar participation of domestic auto manufacturers and energy companies is lacking.

Responses to Recommendations from NRC Phase 4 Review

> ***NRC Phase 4 Recommendation 4-2.*** The fuels pathways integration effort provides strategically important input across different hydrogen pathways and different technical teams to guide U.S. DRIVE Partnership decision making. In this time of budget restraints, the program of the fuel pathway integration technical team should be adequately supported in order to continue providing this important strategic input.
>
> ***Partnership Response.*** The Partnership agrees with the Committee that the Fuel Pathway Integration Technical Team (FPITT) provides strategically important input that informs the Executive Steering Group (ESG) and as well as other technical teams. FPITT activities align with the Systems Analysis subprogram in the U.S. Department of Energy's (DOE) Fuel Cell Technologies Office, which DOE continues to support.

Committee Assessment of Response to 4-2. The FPITT efforts are well coordinated and the results of the analyses are being used to direct R&D efforts to focus on appropriate critical factors and to prioritize the project areas. However, as suggested, this may be taken a step further to make some bold decisions with respect to the ultimate viability of different approaches to update the roadmap and timeline for future transportation scenarios.

Findings and Recommendations

While electricity infrastructure required for the initial introduction of BEVs, PHEVs, and related vehicle technology options already exists, hydrogen infrastructure is practically nonexistent. Therefore, market introduction of HFCVs faces a daunting challenge. Moreover, with ongoing improvements in engine and battery technologies, competition for HFCVs is getting more intense. Nevertheless, HFCVs are being rolled out and made available, especially by automakers not part of U.S. DRIVE, like Toyota and Hyundai. But there is a lack of hydrogen infrastructure, which can derail deployment. Some states, like California, Connecticut, and New York, along with companies like Toyota and Honda, are promoting infrastructure build-out by providing funding.[13] But to date there are very few operating fueling stations to support the projected market for HFCVs. High station cost is an obvious barrier.

Finding 3-28. Technology developments for hydrogen fuel cell vehicles are not enough to bring them to market; there is a critical need for hydrogen infrastructure development.

---

[13] See, for example, the California Hydrogen Business Council at https://www.californiahydrogen.org/content/chbc-releases-report-private-financing-hydrogen-fueling-stations. A 2017 report by the California Energy Commission also addresses the time and cost for 100 hydrogen fueling stations and can be found at http://www.energy.ca.gov/serp.html?q=CEC-600-2017-002&cx=001779225245372747843%3Actr4z8fr3aa&cof=FORID%3A10&ie=UTF-8&submit.x=13&submit.y=9.

Recommendation 3-15. The Executive Steering Group should address issues (e.g., how will fueling stations be installed and by whom, who operates them, who will produce hydrogen, how will investments occur in fueling infrastructure without sufficient fuel cell vehicles on the road and vice versa, etc.) related to hydrogen infrastructure and assess U.S. DRIVE's role to formulate an action plan to address the issues and barriers.

Finding 3-29. Although industrial gas companies currently have the most experience with hydrogen production, delivery, and infrastructure, they are not core members of the Partnership; some of them serve as associate members of technical teams, but not on a consistent basis.

Recommendation 3-16. U.S. DRIVE should consider having industrial gas companies involved in hydrogen infrastructure activities as permanent members rather than as temporary associate members.

SAFETY, CODES, AND STANDARDS

In addition to the importance of developing a hydrogen refueling infrastructure, global codes and standards play a critical role in laying the groundwork for commercialization of hydrogen and fuel cell technologies. It is important to coordinate the SCS efforts with international bodies to harmonize the regulations, codes, and standards. The hydrogen codes and standards technical team (CSTT) of the U.S. DRIVE is focused on "efforts to enable and facilitate appropriate R&D to support technology readiness for the development of safe, performance-based technical codes and standards that support the 2015 commercialization decision for widespread consumer use of fuel cells and hydrogen-based technologies with commercialization by 2020" (U.S. DRIVE, 2013a). Although CSTT efforts are focused on hydrogen, the current membership does not include any industrial gas company; efforts are under way to add one as an associate member. The CSTT coordinates its efforts with those of the other technical teams for consistency and synergy. The CSTT roadmap, updated in June 2013, provides information and detailed discussion of specific R&D, testing, and analysis in the areas of materials compatibility, risk assessment and behavior, accelerated testing, component performance, and fuel quality (U.S. DRIVE, 2013a). CSTT provides support for the development of HFCV standards as well as hydrogen fueling station standards.

The CSTT Roadmap and the R&D priorities identified in it are incorporated in the DOE multiyear research, development, and demonstration plan. The DOE-funded projects to support SCS development, which are considered relevant to the Partnership, are listed in Table 3-14.

The SCS programs at DOE support and facilitate the development and publication of essential codes and standards and of domestic and international regulations. The DOE conducts R&D to provide critical data and information

TABLE 3-14 Department of Energy Projects Related to Hydrogen Codes and Standards

Project ID	Presentation Title	Organization
scs001	National Codes and Standards Deployment and Outreach	National Renewable Energy Laboratory
scs002	Component Standard Research and Development	National Renewable Energy Laboratory
scs004	Hydrogen Safety, Codes and Standards: Sensors	Los Alamos National Laboratory
scs005	Research and Development for Safety, Codes and Standards: Materials and Components Compatibility	Sandia National Laboratories
scs007	Hydrogen Fuel Quality	Los Alamos National Laboratory
scs011	Hydrogen Behavior and Quantitative Risk Assessment	Sandia National Laboratories
scs017	Hands-On Hydrogen Safety Training	Lawrence Livermore National Laboratory
scs019	Hydrogen Safety Panel, Safety Knowledge Tools, and First Responder Training Resources	Pacific Northwest National Laboratory
scs021	National Renewable Energy Laboratory Hydrogen Sensor Testing Laboratory	National Renewable Energy Laboratory
scs022	Fuel Cell & Hydrogen Energy Association Codes and Standards Support	Fuel Cell and Hydrogen Energy Association
scs024	Hydrogen Contaminant Detector	National Renewable Energy Laboratory
scs025	Enabling Hydrogen Infrastructure through Science-Based Codes and Standards	Sandia National Laboratories

SOURCE: Project numbers are from the Department of Energy's 2016 Annual Merit Review, see Cooper (2016a).

needed to define requirements in developing codes and standards. For example, a data-driven science-based approach enabled an update to the hydrogen bulk storage separation distances used in key codes (e.g., National Fire Protection Association [NFPA] codes 2 and 55). As a result, required separation distances were reduced by as much as 50 percent in some instances. Direct DOE support has saved 3 to 5 years in the development of relevant codes and standards, according to Dr. Will James, who heads the efforts at DOE (Green Car Congress, 2016). CSTT achievements during 2012-2016 include support for the development of Global Technical Regulation (GTR), development of critical capability at NREL for component safety and reliability testing, and hydrogen sensor field trials at California fueling stations. In 2015, PNNL developed the Hydrogen Tools Portal[14] for disseminating hydrogen safety knowledge to critical user groups (U.S. DRIVE, 2015). This should be very useful in hydrogen fueling commercialization efforts.

Jesse Schneider (BMW) mentioned at the SAE World Congress (Green Car Congress, 2016) that in addition to automakers teaming up for fuel cell vehicle commercialization, work on infrastructures is under way in three continents: Asia (Japan), North America (United States), and Europe (Green Car Congress, 2016). In 2016, each of these areas plans to have more than 50 stations and hundreds more within the next 5 years. These stations are being standardized to NFPA/International Code Council code and are helped by the ISO Technical Specification 198801, which was published in 2016, giving a baseline of safety and performance.

Significant Barriers and Issues That Need to Be Addressed

While significant progress is being made in the area of SCS for safe use of hydrogen as a fuel, a critical technical barrier to the widespread use of hydrogen is usage and access restrictions, such as in parking structures, tunnels, and other areas. Other remaining barriers include high cost and limited availability of hydrogen components and equipment, and corresponding reliability data needed to develop codes and standards. Other nontechnical barriers related to SCS include limited availability of insurance, lack of knowledge and training for officials, lack of consistency in regulations across states, and limited participation of businesses.

Responses to Recommendations from the NRC Phase 4 Review

NRC Phase 4 Recommendation 2-4. The Partnership should place a much higher priority on the safety, codes, and standards (SCS) program and accelerate the date for final regulations, codes, and standards to 2014. The committee still recommends that, if the budget allows, the scope of the SCS program be expanded to cover all vehicle/fuel combinations being considered by DOE. This would include natural gas, battery electric vehicles, plug-in hybrid electric vehicles, biofuels, and other combinations that are appropriate.

Partnership Response. To clarify, U.S. DRIVE's technical scope does not currently include natural gas or biofuels—partners conduct safety, codes, and standards and other activities related to these pathways through programs outside of the Partnership. Also, as noted in the Partnership's response to Recommendation 2-1 of the Third Review: "vehicle safety is a regulated activity and considered outside of the Partnership scope of precompetitive R&D. Each individual partner places a high priority on safety and maintains its own safety program; partners coordinate with each other on safety issues related to new technologies, as appropriate." The U.S. DRIVE Codes and Standards Technical Team focuses specifically on hydrogen and fuel cell codes and standards. Codes and standards issues related to plug-in electric vehicles are handled through the Grid Interaction Techni-

[14] The website for the Hydrogen Tools Portal is http://h2tools.org, accessed September 7, 2016.

cal Team. U.S. DRIVE partners consider the development and promulgation of key codes and standards needed for the near- and long-term deployment of hydrogen, fuel cell, and plug-in electric vehicles to be a high priority. One example of this is the Global Technical Regulation, which passed in December 2012 in large part due to the technical support from U.S. DRIVE partner organizations. Final dates for regulations, codes, and standards are set by outside organizations and not within the control of U.S. DRIVE partners.

Committee Assessment of Response to 2-4. The Partnership response is appropriate and satisfactory. One agency that controls some of the relevant regulations, and is a member of the team, is DOT. Through DOT, the Partnership should be able to influence acceleration of regulation finalization.

NRC Phase 4 Recommendation 2-5. The Partnership should plan and execute a tank testing program for the Type 3 and Type 4 tanks that are expected to be used in passenger vehicles.

Partnership Response. The Partnership agrees that additional testing on Type 3 and Type 4 tanks is critical. In support of this Partnership priority, the U.S. Department of Energy (DOE) has funded work at Sandia National Laboratory (SNL) to conduct testing and analysis on high-pressure tanks. SNL has completed an analysis and developed a numerical model to investigate tank performance. This phase of testing was conducted at a low range of working pressures, and the next phase will include high pressure testing. DOE also funded round-robin tank testing on Type 4 tanks to determine the proper test measurement protocols. The results of this work will benefit current tank standards and regulations. Future plans include the validation of hydraulic and pneumatic tank cycling testing as well as fire testing by 2015.

Committee Assessment of Response to 2-5. While the response is adequate, and necessary work is being conducted as per the update provided in June 2016, it does not mention follow-up implementation actions that utilize the results of these tests. Also, acceleration of these programs would be helpful.

NRC Phase 4 Recommendation 2-6. As candidate materials are identified, the Partnership should expand the safety, codes, and standards program to identify new contaminants that may be given off by adsorbent or chemical storage systems. These activities should be coordinated with the Storage Systems Center of Excellence at the Savannah River National Laboratory.

Partnership Response. To clarify, through its Fuel Cell Technologies Office, the U.S. Department of Energy (DOE) manages a Safety, Codes, and Standards activity. However, the Partnership's Codes and Standards Technical Team (CSTT) focuses specifically on hydrogen codes and standards, as vehicle safety is a regulated activity and considered outside the U.S. DRIVE scope of precompetitive R&D. Each individual partner places a high priority on safety and maintains its own safety program; partners coordinate with each other on safety issues related to new technologies, as appropriate.

The Partnership agrees with the Committee's recommendation to identify new contaminants that may be given off through adsorbent or chemical hydrogen storage systems, and through the CSTT, continues to monitor the progress of hydrogen storage systems including compressed gas and solid-state materials, e.g., adsorbents and chemical and metal hydrides. The CSTT receives feedback through joint technical team meetings with the Hydrogen Delivery and Hydrogen Storage teams. U.S. DRIVE partners place high priority on the development and promulgation of key codes and standards in advance of near-term deployments, which focus on compressed gas storage systems. In support of this priority, DOE has been actively involved in the development of fuel quality standards, SAE J2719 and ISO 14687-2, that are harmonized and in place for the initial deployment of light-duty fuel cell vehicles. These standards incorporate a science-based approach to provide limits on contaminants that could impact the fuel cell performance. The CSTT will continue to evaluate materials-based storage technologies as they advance.

Committee Assessment of Response to 2-6. Considering the Partnership scope, the response seems appropriate. While the Partnership has worked on the development of fuel quality standards, the committee feels that it would be complementary if the fuel cell team initiates a program to improve tolerance of fuel cells to impurities.

NRC Phase 4 Recommendation 2-7. The Partnership should consider phasing out and turning over to industry for commercialization the stationary H2 sensor effort that has been supported for several years and consider starting a new program on inexpensive H2 sensors for vehicles.

Partnership Response. The referenced hydrogen sensor effort has been funded by the U.S. Department of Energy (DOE). Although DOE believes industry involvement is critical, the sensor technologies that it is funding can be used in a variety of technologies and applications including vehicles, stationary systems, and infrastructure, and should not be phased out at this time. DOE has funded work at Lawrence Livermore (LLNL), Los Alamos National Laboratory (LANL), and Intelligent Optical Systems to develop low-cost, durable, and reliable hydrogen safety sensors. The collaboration between LLNL and LANL has developed pre-commercial prototypes using a solid-state electrochemical sensing technology for hydrogen safety applications. LLNL and LANL are also working with ESL ElectroScience on sensor design and processing consideration for commercialization of the technology. In addition, the DOE supports the National Renewable Energy Laboratory's hydrogen sensor laboratory, which provides critical services and support to stakeholders in the hydrogen community including end-users, sensor developers, and regulatory agencies. These efforts entail a variety of approaches that will improve the reliability and reduce the cost of the sensors.

Committee Assessment of Response to 2-7. As per the update provided in June 2016, DOE has ended the funding on the referenced safety sensor. The LANL/LLNL solid-state electrochemical sensor is now ready for commercialization, and DOE continues to fund NREL's sensor validation laboratory to support industry.

NRC Phase 4 Recommendation 2-8. The Partnership should consider starting a hydrogen vehicle emergency response R&D effort similar to that now being conducted for electric and plug-in vehicles. One of the issues that should be studied is how to depressurize a damaged tank.

Partnership Response. As noted in responses to Recommendations 2-4 and 2-6, hydrogen vehicle safety is not addressed, explicitly, under U.S. DRIVE. Individual partners manage their own safety programs internally. However, the U.S. Department of Energy (DOE) maintains an active hydrogen safety effort and coordinates closely with U.S. DRIVE partner organizations and others. DOE has already started a hydrogen emergency response effort, recognizing the importance of an emergency response R&D program to promote the use of technical data in the development of training for first responders. Recently, an SAE international committee started an effort to develop an industry standard for hydrogen/fuel cell vehicle emergency response for first and second responders. The standard will address the potential consequences associated with hydrogen vehicle incidents and suggest common procedures to help protect emergency responders, tow and/or recovery, storage, repair, and salvage personnel after an incident has occurred. In addition, strong collaboration already exists between the U.S. DRIVE partners, including DOE, and the National Fire Academy, as well as other training organizations and agencies to develop training manuals, technical assistance resources, and relevant safety training materials. Examples include the Introduction to Hydrogen Safety for First Responders, the Hydrogen Incident Reporting Database, and deployment of advanced level prop-based course for first responders. These materials draw from information in the Partnership's Codes and Standards Technical Team Roadmap, which addresses issues related to hydrogen behavior, hydrogen-fueled vehicles, fueling infrastructure, and the fuel-vehicle interface. Within the DOE hydrogen safety program, the Partnership's RD&D effort has contributed to material developed in the training of first responders today.

Committee Assessment of Response to 2-8. The Partnership's response is satisfactory. The issue of depressurizing a damaged tank identified in the NRC recommendation is important, and CSTT should continue to work with appropriate organizations to support relevant activities within their scope.

Findings and Recommendations

Finding 3-30. While testing and analysis of high-pressure tanks is progressing well within the DOE-funded program, there is a need to accelerate these activities. Also there is no mention in the codes and standards technical team's plan for follow-up implementation actions that utilize the results of these tests.

Recommendation 3-17. The codes and standards technical team should assess the results of high-pressure tank testing and prepare a plan for implementation actions that utilize the results of testing.

Finding 3-31. While the Partnership has worked on the development of a fuel quality standard, it is also apparent that high-quality hydrogen leads to higher cost of hydrogen. One potential approach to avoid this cost escalation is to improve the tolerance of fuel cells to impurities.

Recommendation 3-18. U.S. DRIVE should explore the possibility of improving fuel cell tolerance to impurities in hydrogen so that a lower-quality, and hence lower-cost, hydrogen may be used.

ELECTRIC DRIVE SYSTEMS AND POWER ELECTRONICS

Background

Under global environmental pressures to reduce tailpipe emissions and dependency on petroleum as a source of energy for ground transportation, most automotive OEMs have been working for decades on the development of electric and hybrid power train systems with zero or ultra-low tailpipe emissions, respectively. Recent advances in battery and electric drive technologies made it possible for vehicle manufacturers to commercially deploy electrified hybrid and electric vehicles (cars, trucks, and buses). Despite continued environmental pressures and evidence of global warming, market penetration of electrified vehicles is growing at a slow rate for a variety of reasons, including cost, electric-only range, and battery charging time, in addition to a noticeably weak consumer acceptance of the technology. The electric drive system (consisting of an electric motor and an electronic controller) is a critical part of electrified power trains for light-duty vehicles. Therefore, a key objective of the U.S. DRIVE Partnership is the development of technologies addressing the electric drive components' cost, weight, and size to help expedite electrified power train market penetration. The Partnership has established cost and technical targets for the electric traction system for 2010, 2015, and 2020 as listed in Table 3-15.

The DOE budgets for this electric drive technology R&D were approximately \$24 million for FY 2014, \$21 million for FY 2015, and \$38.1 million in FY 2016.

A variety of electric drive technologies have been investigated over the years by the industry, academia, and government laboratories, including brush and brushless direct current (DC) and alternating current (AC) motors as they vary in their performance, reliability, and cost. Because of the critical impact of the drive efficiency and power density (weight) on the electric range of electrified vehicles, the leading contenders have been the following: (1) brushless permanent magnet DC motor (BLDC); (2) induction motor (IM); (3) switched reluctance motor (SRM); and (4) synchronous reluctance motor (SyRM). The brushless permanent magnet motor using rare earth (RE; NdFeB) magnets has been the choice for many OEMs (GM, Ford, Toyota, Nissan, Honda) because of

TABLE 3-15 Electric Traction System Technical Targets

	2010	2015	2020
Cost (\$/kW) (at 100,000 units/year)	<19	<12	<8
Gravimetric power density (kW/kg)	>1.06	>1.2	>1.4
Volumetric power density (kW/L)	>2.6	>3.5	>4
System efficiency (%)	>90	>93	>94

SOURCE: Data from El-Refaie (2016).

its superior performance (efficiency, power density, and torque density), which enabled them to reach a reasonable commercial range despite the limited energy density offered by available lithium-ion batteries. Only a few, including Tesla Motors and some conversion companies, opted for IM technology, as a low-risk choice, since induction motors have served as the workhorse of the industry for decades. Furthermore, methods for induction motor control for operation from a battery source (DC) are known and have also been established for a long time. The switched reluctance motor offers great advantages in reliability and cost but suffers from an inherent audible noise problem that researchers have not been able to resolve to the level required for commercial implementation (Omekanda, 2013).

However, the preceding analysis and situation have been disrupted by two factors: (1) The cost of rare earth magnets skyrocketed a few years ago, sending shock waves across the globe to users and causing deep concerns about long-term availability. This has prompted most users to look for alternatives. (2) Battery technologies have made significant progress and battery energy densities showed considerable improvements, so a slightly lower motor efficiency or power density is less critical. Therefore, the focus of the U.S. DRIVE electric drive R&D activity is to develop technologies aimed at the reduction or elimination of rare earth magnets in motors, and simultaneously, the adoption of low-loss wide bandgap (WBG) semiconductors to further boost electric drive system efficiency (Dawsey and Rogers, 2016). The effort to eliminate RE magnets is exploring two possibilities: (1) maintaining the Brushless Permanent Magnet (PM) motor type, but developing advanced high-energy non-rare earth magnets (AlNiCo, Ferrite, or dysprosium-free RE magnets) to replace the NdFeB magnets having the super-expensive Dysprosium content currently used; and (2) reconsidering other non-PM motor types (IM, SRM, SyRM, wound field excited motors) but incorporating innovative structures/assemblies, and effective thermal and noise management techniques as lower-cost alternatives.

Current Status

Progress was made by the Partnership both on the research and the development fronts pushing the system cost and performance closer to the set targets listed in Table 3-16 (U.S. DRIVE, 2013b).

Estimating drive system performance and cost based on promising motor concepts (projects edt044 and edt045 listed in Table 3-17) and advanced inverters (project edt040 in Table 3-17) that have been evaluated point to a drive system that can meet the 2015 set targets, with potential for further improvements toward the 2020 targets.

TABLE 3-16 U.S. DRIVE Electric Drive Targets

		U.S. DRIVE 2015 Target			U.S. DRIVE 2020 Target		
Parameter	Units	Motor	Power Electronics	Total Drive System	Motor	Power Electronics	Total Drive System
Power density (gravimetric)	kW/kg	1.3	12	1.2	1.6	14.1	1.4
Power density (volumetric)	kW/L	5	12	3.5	5.7	13.4	4
Efficiency	%	—	—	>93	—	—	>94
Specific cost (at 100,000 units/year)	$/kW	7	5	12	4.7	3.3	8

SOURCE: U.S. DRIVE (2013b).

TABLE 3-17 Department of Energy Electric Drive Projects Related to U.S. DRIVE Goals and Targets

Project ID	Presentation Title	Organization
edt006	Benchmarking EV and HEV Technologies	Oak Ridge National Laboratory
edt015	DREAM (Development of Radically Enhanced Alnico Magnets)	Ames
edt032	North American Electric Traction Drive Supply Chain Analysis: Focus on Motors	Synthesis Partners
edt040	Next-Generation Inverter	General Motors
edt044	Unique Lanthide-Free Motor Construction	UQM Technologies, Inc.
edt045	Alternative High-Performance Motors with Non-Rare Earth Materials	General Electric
edt049	Advanced Packaging Technologies and Designs	Oak Ridge National Laboratory
edt053	Electric Drive Inverter Research and Development	Oak Ridge National Laboratory
edt054	Innovative Technologies for Converters and Chargers	Oak Ridge National Laboratory
edt058	Advanced Low-Cost SiC and GaN Wide Bandgap Inverters for Under-the-Hood Electric Vehicle Traction Drives	APEI, Inc.

TABLE 3-17 Continued

Project ID	Presentation Title	Organization
edt059	High-Temperature DC Bus Capacitors Cost Reduction and Performance Improvements	Sigma Technologies International
edt060	High-Performance DC Bus Film Capacitor	General Electric
edt061	Cost-Effective Fabrication of High-Temperature Ceramic Capacitors for Power Inverters	Argonne National Laboratory
edt062	Non-Rare-Earth Motor Development	Oak Ridge National Laboratory
edt063	Performance and Reliability of Bonded Interfaces for High-Temperature Packaging	National Renewable Energy Laboratory
edt064	Electric Motor Thermal Management Research and Development	National Renewable Energy Laboratory
edt065	Brushless and Permanent Magnet Free Wound Field Synchronous Motor (WFSM)	University of Wisconsin, Madison
edt067	High-Efficiency High-Density GaN-Based 6.6 kW Bidirectional On-Board Charger for PEVs	Delta Products Corporation
edt068	Gate Driver Optimization for WBG Applications	Oak Ridge National Laboratory
edt069	Power Electronics Thermal Management Research and Development	National Renewable Energy Laboratory
edt070	Thermal Performance Benchmarking	National Renewable Energy Laboratory
edt072	30 kW Modular DC-DC System Using Superjunction MOSFETs	University of Colorado
edt073	Evaluation of an APEI 88 kW SiC Inverter with Next-Generation Cree 900 V SiC MOSFET Technology for Ford Automotive Systems	Cree

SOURCE: Project numbers are from the Department of Energy's 2016 Annual Merit Review, see Cooper (2016a).

Assessment of Progress and Key Achievements

The U.S. DRIVE focus on electric drive technologies is supported through projects funded by the DOE VTO, which, as noted previously has a budget of $38.1 million in FY 2016. Table 3-17 lists DOE projects related to meeting U.S. DRIVE's goals and targets in the electric drive area.

Electric Motors

Excellent progress has been made in assessing various motor topologies and advanced materials:

- In project number edt044, UQM has developed, built, and tested a new PM motor with an innovative magnetic design and incorporating new non-RE magnets (AlNiCo) developed by Ames Lab. The proof-of-concept (POC) motor demonstrated performance very close to requirements with off-the-shelf magnet material. The work on increasing magnet properties looks promising and would reduce magnet content and cost. Motor build will demonstrate the feasibility of the approach to meet or exceed the 2020 DOE requirements with Ames-designed magnets and optimized cooling methods from NREL (Ley, 2016).
- In project number edt045, GE has completed the experimental evaluation of several proof-of-concept motors: (1) with reduced RE (Dy-free flux switching type), (2) non-RE (Spoke-Ferrite), (3) no-magnet (DC-biased switched reluctance-SRM), and (4) SyRM using a special magnetic laminate patterned with nonmagnetic regions and a new high-temperature slot liner material. Results of the work of item 4 on the SyRM look most promising and seem to approach the 2020 DOE set targets (El-Refaie, 2016).
- In project numbers edt062 and edt065, the ORNL and its partners have also studied motors without rare earth permanent magnets—namely, Concentrated and distributive wound ferrite machines, brushless field excited (BFE), and synchronous reluctance (SyRM)—while using advanced soft magnetic materials with reduced magnetic loss to enhance performance (Burress, 2016; Ludois, 2015). The simulation results were encouraging. The surprising result is that two distributive wound ferrite machine designs are expected to reach DOE 2020 motor targets and have similar performance to rare earth PM motors despite the large difference in energy product, which is a measure of magnet strength, between ferrite and RE magnets (RE magnets are 5-10 times better than ferrite). The details of the designs were not revealed at the June 22-23, 2016, committee meeting due to pending patents.
- Most exciting results came from the ORNL's work on a multi speed motor concept (Dawsey and Rogers, 2016), which uses electronic switching to reconfigure the stator windings at three different motor speeds potentially resulting in a 24 percent reduction in drive cycle loss (Dawsey and Rogers, 2016). While the concept is not new, it is the ability to use only a small number of power switches to accomplish the switching that is innovative. If successful this approach will help boost motor efficiency and reduce cooling needs, thus reducing motor size and weight.

Power Electronics

Significant progress was reported by the Partnership in power electronics. This was achieved by using innovative packaging and integration of classic inverters and converter configurations and also by exploring the use of WBG devices for automotive power electronic systems. Key achievements are summarized below:

- In project number edt040, which was completed in December 2015, General Motors in collaboration with ORNL, NREL, and suppliers has developed and tested its next-generation inverter, demonstrating a capability for scalability, high efficiency, and high power density, which is projected to meet all U.S. DRIVE 2020 targets for performance and cost. This was achieved by an innovative design that integrates active components and reduces/eliminates supporting components (Zhao, 2016).
- Many projects aimed at increasing the level of understanding of WBG devices, while others focused on developing high-frequency circuitry and high-temperature components to sustain the high-temperature and high-switching frequency environment enabled by the WBG devices. These include WBG device characterization, as in project number edt53, as well as evaluating converters and inverters using WBG devices (Chinthavali, 2016). In project number edt53, a 55-kW traction drive WBG silicon carbide (SiC)-based inverter was designed and will be built for testing in 2017. Another WBG traction inverter investigation was conducted by APEI and Cree in project numbers edt58 and edt73, leading to necessary fundamental learning (Olejniczak, 2016; Casady, 2016). Also, in project number edt67, Delta Products is evaluating a gallium nitride (GaN)-based onboard charger (Zhu, 2016). An integrated SiC-based onboard charger and DC-DC converter demonstrates double the power density with a 40 percent reduction in cost compared to the state-of-the-art non-WBG silicon-devices (Dawsey and Rogers, 2016).
- NREL has demonstrated ribbon electrical interconnects to replace the current round wire interconnects, which enables higher power densities and lower parasitic inductances that are particularly critical for WBG devices (Dawsey and Rogers, 2016). Ribbon interconnects of various geometries and materials were tested under elevated temperature, temperature cycling (thermal shock), corrosion and highly accelerated life test (combined vibration and thermal cycles) and exhibited similar reliability to wire interconnects under these conditions.
- Thermal stack-up enables full potential of WBG Devices. New Sintered-Silver (Ag) Interconnects technology, project number edt063, developed by ORNL, is showing promise and is under intense testing to understand and demonstrate the ability to consistently and effectively create reliable sintered joints (DeVoto, 2016).

Barriers and Issues That Need to Be Addressed

There are two key barriers to closing the gaps to Partnership targets: (1) component cost, particularly that of rare earth permanent magnets, which, in some cases, can be more than half the motor cost; and (2) the high power loss associated with electronics, which leads to heavy and expensive cooling systems (U.S. DRIVE, 2013b). By improving the total drive system efficiency, the thermal management system can be reduced leading to overall reductions in volume, weight and cost.

Response to Phase 4 Recommendations

> ***NRC Phase 4 Recommendation 3-14.*** The U.S. DRIVE Partnership should leverage the various investigations on Wide Bandgap materials such as silicon carbide (SiC) and should determine how best to utilize these in power electronics for vehicular application.
>
> ***Partnership Response.*** The Advanced Power Electronics and Electric Motors (APEEM) program has a comprehensive approach to increasing performance and efficiency while reducing costs of power electronic systems in an effort to meet U.S. DRIVE targets. APEEM activities are ongoing to leverage WBG related research and development within the program and partnership, as well as other DOE programs and Federal agencies, to determine how to best utilize WBG devices in power electronics for vehicle applications. The APEEM efforts with industry (Delphi and General Motors) and the National Laboratories include focused research on wide bandgap (WBG)-based power electronics systems for automotive applications. Ongoing benchmarking of WBG devices provides data and analysis for state of the art technologies. WBG material, specifically silicon carbide (SiC) and gallium nitride (GaN), offer substantial improvements in efficiency as well as reduced size and weight when compared to silicon (Si) based systems currently being used in automotive applications. These applications require high efficiency at high temperatures making WBGs best suited to meet automotive requirements. Higher temperature operation enables cost and weight savings through reduced size heat sinks and packaging requirements and the potential to eliminate secondary cooling loops. With increased switching frequency, passive components can be minimized resulting in further reductions in cost, weight and volume. The APEEM program is utilizing expertise and advanced technologies from ongoing efforts to support a WBG inverter prototype with advanced controls by leveraging recent R&D including air-cooled traction drive designs, high frequency dc-dc converters with reduced passives, charging applications, and assessment of WBG technology devices. In addition there is ongoing collaboration with other DOE programs and Federal agencies (through the Interagency Advanced Power Group).

Committee Assessment of Response to 3-14. The committee appreciates the intense activities and focus of the DOE-APEEM program on WBG devices aiming at improving the efficiency and reducing the size and weight of the power electronic system. It is expected, however, that through these investigations one could determine which of the WBG devices (GaN versus SiC) is the more appropriate and cost effective for automotive applications and possibly redirect efforts to accelerate its readiness for implementation.

NRC Phase 4 Recommendation 3-15. *The U.S. DRIVE Partnership should determine the potential and limitations of designing motors with permanent-magnet materials using less rare (RE) earth metal.*

Partnership Response. The Advanced Power Electronics and Electric Motors (APEEM) program is taking a comprehensive approach to addressing the limitations of designing motors with permanent-magnet materials using less rare earth metal. High-energy rare earth material price and its consequent limitation on meeting U.S.DRIVE targets is the focus of ongoing R&D activities. The APEEM portfolio includes focused research on magnet material and processes in the Ames National Laboratory Beyond Rare Earth Magnets (BREM) program plus motor design and material research under way in the national laboratories and industry programs (General Electric, UQM Technologies) on traction motor development to eliminate or significantly reduce the use of rare earth materials in magnets. The existing APEEM project portfolio is complemented by ARPA-E industry funding on rare earth alternatives in critical technologies (REACT) motor development activities. These activities recognize that today's hybrid and plug-in electric vehicles rely almost exclusively on interior permanent magnet electric motors designed with high energy rare earth magnets. The real limitation of using less or alternative permanent magnet materials is lower specific power and power density that in turn inhibit their packaging flexibility in the vehicle. However, there is little opportunity for using less RE material based on assessment of commercial interior magnet traction motor performance over the past decade showing that permanent magnet content is optimized to 77kW/kg. Vehicle integration will be seriously limited by larger size traction motors. Therefore, the APEEM programs are technically focused, results oriented activities that fund industry and the national laboratories to develop advanced alternative magnetic materials and motor technologies with the potential to significantly reduce the high cost associated with existing rare earth magnets.

Committee Assessment of Response to 3-15. The committee agrees that the APEEM programs are designed to address the issues associated with using magnets with less or no RE material. The concern is that much of the focus was directed toward developing new AlNiCo or ferrite material with improved performance but still far below that of NdFeB magnets, which as the preceding response indicates would make it difficult to meet all the DOE targets. The approach followed by Honda (Williams, 2016), which appears to be successful, is to eliminate the need for only the super-expensive RE additives (dysprosium) used for protecting the magnet against demagnetization at high temperatures, by developing a new manufacturing process that gives the high-energy NdFeB magnet a natural protection at high temperature.

NRC Phase 4 Recommendation 3-16. The U.S. DRIVE Partnership should make a comprehensive assessment of the various methods available (some of these are discussed in the section titled "Thermal Management" in this chapter) to reduce the thermal resistance between the chip and the heat sink and establish their relative value to existing techniques in production vehicles.

Partnership Response. The APEEM Program assesses state-of-the-art thermal management technologies and funds ongoing R&D activities to develop advanced technologies to reduce the thermal resistance in power electronics, motor packaging, and heat exchangers. Research emphasizes analysis and development of advanced interfaces, interconnects and materials with low thermal resistance and improved reliability for power electronics

> and motor packaging. Bonded interface materials with a thermal resistance reduction of a factor of 10 (for a thickness of 75 μm), over materials used in production vehicles such as the 2007 Toyota Camry, have been demonstrated. Some of these bonded materials are also demonstrating good reliability. Alternative power electronics packaging configurations with lower thermal resistance are being investigated. To reduce the heat exchanger resistance, different techniques including air cooling, single-phase liquid cooling, two-phase cooling, heat transfer surface enhancements, as well as efficient heat spreading techniques are being assessed. Some of the advanced liquid cooling and two-phase cooling techniques investigated in the program have demonstrated the potential to reduce thermal resistance, and consequently increase the power electronics power-per-die-area as well as power density by 60% to 100% in comparison to the 2008 Lexus LS 600h power module. Modeling efforts have also demonstrated the potential to reduce overall thermal resistance and increase the motor power density by up to 100% over state-of-the-art technique. Ongoing thermal management R&D to reduce thermal resistance will yield techniques and technologies to meet inverter, motor and electric traction drive system-level APEEM Program targets for cost, power density, specific power and reliability.

Committee Assessment of Response to 3-16. The committee agrees that the current DOE programs follow the above recommendation with a potential for a positive impact.

Appropriate Federal Role

Despite continued environmental pressures and evidence of global warming, market penetration of electrified vehicles is growing at a slow rate partially due to cost, size, and weight (impacting range) of the electric drive, which is the core of all electrified power trains. There are, however, fundamental challenges in dealing with issues of size, weight, and cost of the electric drive, discussed in this report, which require a concerted national effort to address. Addressing issues such as finding a replacement for the costly rare earth magnets used in the motor or finding means to produce WBG devices at affordable cost would require resources beyond the capabilities of individual OEMs and are likely to exist at the U.S. national laboratories. Therefore, in the committee's view, the support of the federal government in forming this Partnership and involving the national laboratories is essential for a successful outcome, which would not only expedite electrified power train market penetration and help with national competitiveness but would also benefit other industries of national interest (e.g., defense).

Findings and Recommendations

It is the committee's opinion that the APEEM project portfolio pursued by the U.S. DRIVE Partnership, and the supporting DOE-led projects in the motors and power electronics areas, are appropriate to address cost, size, and weight challenges. Several motor configurations and design variations are under investigation to address the high cost of RE magnets. Significant progress was made toward meeting the Partnership targets in the power electronics area, achieved through innovative packaging and integration of inverter and converter components with the use of WBG devices (GaN and SiC). Given the inherent cost advantage of GaN devices grown on Si-substrate compared with SiC on SiC substrate (due to the much higher cost of SiC compared to Si), it is expected that GaN will ultimately be the winner among these two competing technologies for automotive applications. Historically, SiC devices have been the focus of research for many years, as they possess higher voltage and temperature capabilities than GaN devices. Operating at these high levels of temperatures requires other circuit components to be also capable of these temperatures, which is cost prohibitive for automotive applications but not so for other cost-tolerant applications, such as for defense.

In addition to meeting the DOE targets for cost and power density, propulsion drives must meet other criteria of automotive applications, such as electromagnetic interference, audible noise level, and instant torque availability. For example, switched reluctance and PM flux-switching types are known to exhibit a level of audible noise beyond what would be acceptable for automotive use. Also, wound-field excited machines exhibit a time delay in torque generation as it takes time for the excitation current and the magnetic field it produces to build up to the level required for full torque generation. Automotive propulsion requires instant torque availability upon demand.

An important element of the DOE VTO project management process is the go/no-go decision used to downselect from many considered options to the few that hold the most promise for meeting program objectives. For example, both ORNL and General Electric have programs that consider a wide range of motor technologies. The ORNL go/no-go decision point follows a thorough and extensive analysis phase, while General Electric carries many technologies to the build and test phase before a go/no-go decision. Considering the availability of accurate analytical tools for almost all motor technologies, it is recommended that DOE follow the excellent and cost-effective ORNL approach, mentioned above as an example, to eliminate options that could not meet fundamental automotive requirements stated earlier.

Finding 3-32. Some of the considered motor technology options, mentioned earlier, have been pursued for an extended length of time despite known inherent limitations. They could have been eliminated earlier on the basis of analysis.

Recommendation 3-19. The U.S. DRIVE Partnership should take into account those specific criteria of the automotive propulsion applications such as electromagnetic interference, audible noise level, and instant torque availability, while searching for the best solution to meet the Department of Energy-set targets for electric drive.

- The U.S. DRIVE Partnership should assess the audible noise produced by some of the motor configurations under consideration (e.g., switched reluctance, PM flux switching types) and determine if it would be commercially acceptable. If noise is found to be truly unacceptable, as anticipated, work on these machine types should be terminated and resources devoted to more promising configurations.

- If wound-field excited machine technology is still under consideration (El-Refaie, 2016), the U.S. DRIVE Partnership should address the field build-up delay in this type of motor, as torque is expected to be available instantly in automotive propulsion applications.

Finding 3-33. Only a few projects are exploring GaN, with the majority focusing on SiC. Given GaN's potential cost advantage, it is expected to ultimately be the preferred choice for automotive applications.

Recommendation 3-20. The U.S DRIVE Partnership should increase the focus on the advancement of gallium nitride technology in order to accelerate its readiness for commercial implementation.

Finding 3-34. There is a tendency within the automotive industry to maintain similar gear-up ratios in order to avoid the additional cost associated with redesigning the transmission for higher motor speeds. Operating the motor at a higher speed will reduce the size, weight, and cost of all its active components including the magnets. However, there are major cost, noise, and reliability challenges to increasing the motor speed, which could be difficult to resolve.

Recommendation 3-21. The U.S. DRIVE Partnership should explore new and innovative alternative gearing concepts, as operating at higher motor speeds translates directly to a smaller and lighter motor, with reduced material cost including magnet cost, which might offset the increased cost of gearing. A trade-off study to assess the total impact on the cost and weight of the total propulsion drive including the gearing system should be undertaken.

ELECTROCHEMICAL ENERGY STORAGE

Background

As noted in Chapter 1 and in the Partnership Plan (U.S. DRIVE, 2016), the overall mission of the U.S. DRIVE Partnership is to "accelerate the development of precompetitive and innovative technologies to enable a full range of efficient and advanced light-duty vehicles, as well as related energy infrastructure." Electrochemical energy storage is a critical component for these advanced vehicles. Electrochemical energy storage (including batteries and supercapacitors) is used in all electric drive vehicles including HEVs, PHEVs and BEVs, and HFCVs. (Note that PHEVs and BEVs are both plug-in electric vehicles that can recharge their batteries from the electric grid.) The electric system for plug-in electric vehicles and HEVs operates at a high voltage, ranging from 200 to 400 V. Recently there is renewed interest in lower voltage systems of 12 and 48 V for start-stop (automatically shuts the internal combustion engine off when the car is idling and restarts when the foot is off the brake) and mild hybrid vehicles (do not have all the functionality of a full HEV).

The Toyota Prius, the first commercial vehicle to employ significant energy storage in an HEV, was introduced in 1999 in the United States. In 2015 there were over 50 models of HEVs available yet they represented only about 2 percent of new light-duty vehicle sales of the automotive market (DOE, 2016a). Initially HEVs used nickel metal-hydride batteries; however, the newer models are incorporating lithium-ion battery technology. Substantial performance improvement and the lower cost of lithium-ion battery technology have also enabled the successful introduction of BEVs and PHEVs in addition to higher-performing HEVs. In 2015 there were 1.26 million plug-in electric vehicles on the road worldwide, over 400,000 of which are in the United States, yet they represented only about 0.7 percent of new light-duty vehicle sales of the automotive market in the United States in 2015 (IEA, 2016; DOE, 2016b). Thus, there is much room for improvement in battery performance and cost reduction to increase market penetration of electric drive vehicles.

High cost remains the main impediment to significant market penetration of plug-in electric vehicles that utilize large batteries. There is also a need to improve battery performance characteristics—that is, energy density, specific energy, operation at extreme temperatures, charging and discharging rates, cycle, and calendar life. These improvements in performance and cost reduction need to be realized while addressing the inherent safety issues associated with lithium batteries, particularly as battery size increases. Lithium-ion battery performance, cost, and safety are being addressed by the DOE and other government entities, automotive OEMs, and battery manufacturers.

The DOE VTO has organized the electrochemical energy storage technology program into five subactivities:

1. Battery technology development;
2. Applied battery research;
3. Battery materials research;
4. Battery testing, analysis, and design; and
5. Manufacturing and process development.

Subactivity (1), battery technology development, efforts are directed toward support of the domestic battery industry to develop battery module and system hardware and related activities. The VTO works closely with the U.S. Advanced Battery Consortium (USABC) in support of cost-shared contracts with developers. In addition, it also directly supports battery and materials suppliers via contracts administered by the National Energy Technology Laboratory aimed at BEV, PHEV, and HEV applications. Subactivity (2), applied battery research, is directed by the national laboratories with ANL in the lead role. It focuses on the next generation of high-energy batteries, optimizing systems incorporating new battery materials and diagnosing and mitigating issues that impact battery performance and life. Subactivity (3), battery materials research, is involved in the development of newer materials and electrochemical couples and the fundamental understanding of specific electrochemical systems for lithium batteries. It also studies electrochemical systems beyond lithium ion with a potential for higher energy and power. This exploratory work is directed by LBNL and conducted by many academic groups with participation of national laboratories and industrial research groups. Subactivity (4), battery testing, analysis, and design, is complementary to battery technology development (1) and is involved in life and abuse testing of deliverable cells and batteries, benchmarking systems from industry, test procedure development, modeling, and battery materials recycling. Subactivity (5), manufacturing and process development, is complementary to applied battery research (2) and is involved in development of kilogram-scale advanced materials and customized scaled processes. It is run by the national laboratories with industry partnerships to develop novel electrode/cell manufacturing technologies.

Individual DOE projects in support of the various subactivities (1)-(5) are listed in Table 3-18. The projects are sorted by the energy storage subactivity and were collated from the 2016 VTO AMR presentations.

In addition, the VTO supports several SBIR contracts to develop new ideas and concepts. As discussed in NRC (2013), longer-term energy storage objectives are also being pursued by DOE-funded R&D in the ARPA-E organization that is aimed at high-risk, game-changing technologies beyond the projected capabilities of lithium-ion batteries. Fundamental R&D activities on materials are also pursued through the DOE BES. The VTO coordinates efforts with other government agencies: the DOE Office of Electricity Delivery and Energy Reliability; the Chemical Working Group of the Interagency Advanced Power Group; the Department of Transportation/National Highway Traffic Safety Administration (DOT/NHTSA); the EPA; and the United Nations Working Group on Battery Shipment Requirements. In addition, it includes interactions with the International Energy Agency's (IEA's) Implementing Agreement on Hybrid Electric Vehicles (IA-HEV), the eight-nation Electric Vehicle Initiative (EVI), and the Clean Energy Research Center (CERC).

While the U.S. DRIVE Partnership does not have a budget or program per se, it provides goals, targets, and roadmaps via the electrochemical energy storage technical team (EESTT) to those entities that do have budgets—for example, the DOE through the VTO, the ARPA-E, SBIR program, and BES. The DOE VTO largely provides funding that supports U.S. DRIVE goals. The USABC establishes its own goals and targets, which are closely coordinated and aligned with U.S. DRIVE, and funds projects that address these. The DOE VTO FY 2016 budget

TABLE 3-18 Department of Energy's Energy Storage Projects Supporting the U.S. DRIVE Mission

EST Subactivity	Project ID	Project Name	Organization
1	ES210	Advanced High Energy Li-Ion Cell for PHEV and EV Applications	3M
1	ES241	Advanced High-Performance Batteries for Electric Vehicle (EV) Applications	Amprius
1	ES289	Advanced Polyolefin Separators for Li-Ion Batteries Used in Vehicle Applications	Entek
1	ES247	High-Energy Lithium Batteries for Electric Vehicles	Envia Systems
1	ES249	A 12V Start-Stop Li Polymer Battery Pack	LG Chem Power
1	ES251	Development of Advanced High-Performance Batteries for 12V Start Stop Vehicle Applications	Maxwell
1	ES265	UV Curable Binder Technology to Reduce Manufacturing Cost and Improve Performance of LiB Electrodes	Miltec UV International
1	ES238	Low-Cost, High-Capacity Lithium Ion Batteries through Modified Surface and Microstructure	Navitas Systems
1	ES267	Commercially Scalable Process to Fabricate Porous Silicon	Navitas Systems
1	ES290	Hybrid Electrolytes for PHEV Applications	NOHMs Technologies
1	ES212	High-Energy, Long Cycle Life Lithium-ion Batteries for EV Applications	Pennsylvania State University (Penn State)
1	ES288	Construction of High-Energy Density Batteries	Physical Sciences Inc.
1	ES239	Scale-Up of Low-Cost Encapsulation Technologies for High-Capacity and High-Voltage Electrode Powders	PneumatiCoat Technologies
1	ES291	SAFT-USABC 12V Start-Stop Phase II	Saft
1	ES240	High-Energy Anode Material Development for Li-Ion Batteries	Sinode Systems
1	ES292	Development of Advanced High-Performance Electrolytes for Lithium-Ion Used in Vehicle Applications	Soulbrain
1	ES293	A Closed Loop Process for the End-of-Life Electric Vehicle Li-ion Batteries	WPI

continued

TABLE 3-18 Continued

EST Subactivity	Project ID	Project Name	Organization
1	ES237	Low-Cost, High-Energy Si/Graphene Anodes for Li-Ion Batteries	XG Sciences
2	ES028	Materials Benchmarking Activities for CAMP Facility	ANL
2	ES030	Cell Analysis, Modeling, and Prototyping (CAMP) Facility Research Activities	ANL
2	ES166	Post-Test Analysis of Lithium-Ion Battery Materials	ANL
2	ES208	New High-Energy Electrochemical Couple for Automotive Applications	ANL
2	ES252	Enabling High-Energy, High-Voltage Li-Ion Cells for Transportation Applications: Modeling and Analysis	ANL
2	ES253	Enabling High-Energy, High-Voltage Li-Ion Cells for Transportation Applications: Project Overview	ANL
2	ES254	Enabling High-Energy, High-Voltage Li-Ion Cells for Transportation Applications: Materials Characterization	ANL
2	ES261	Next Generation Anodes for Lithium-ion Batteries: Overview	ANL
2	ES211	High-Energy Lithium Batteries for PHEV Applications	Envia
2	ES213	High-Energy Density Li-ion Cells for EVs Based on Novel, High-Voltage Cathode Material Systems	Farasis
2	ES262	Next-Generation Anodes for Li-Ion Batteries: Fundamental Studies of Si-C Model Systems	LBNL
2	ES271	New Advanced Stable Electrolytes for High-Voltage Electrochemical Energy Storage	Silatronix
2	ES209	High-Energy, High-Power Battery Exceeding PHEV-40 Requirements	TIAX
3	ES049	Tailoring Spinel Electrodes for High-Capacity Li-Ion Cells	ANL
3	ES235	Characterization Studies of High-Capacity Composite Electrode Structures	ANL
3	ES280	Novel Chemistry: Lithium Selenium and Selenium Sulfur Couple	ANL

TABLE 3-18 Continued

EST Subactivity	Project ID	Project Name	Organization
3	ES286	Development of Novel Electrolytes and Catalysts for Li-Air Batteries	ANL
3	ES231	High Energy Density Lithium Battery	Binghamton University-SUNY
3	ES220	Predicting Microstructure and Performance for Optimal Cell Fabrication	Brigham Young University
3	ES059	Advanced In Situ Diagnostic Techniques for Battery Materials	BNL
3	ES183	In Situ Solvothermal Synthesis of Novel High-Capacity Cathodes	BNL
3	ES281	Multi-Functional Cathode Additives for Li-S Battery Technology	BNL
3	ES287	Exploratory Studies of Novel Sodium-Ion Battery Systems	BNL
3	ES221	A Combined Experimental and Modeling Approach for the Design of High Coulombic Efficiency Si Electrodes	General Motors
3	ES222	Development of Si-Composite Anode for Large-Format Li-ion Batteries	Hydro Quebec
3	ES052	Design of High-Performance, High-Energy Cathode Materials	LBNL
3	ES054	First Principles Calculations of Existing and Novel Electrode Materials	LBNL
3	ES085	Interfacial Processes in EES Systems Advanced Diagnostics	LBNL
3	ES091	Predicting and Understanding Novel Electrode Materials from First Principles	LBNL
3	ES223	Hierarchical Assembly of Inorganic/Organic Hybrid Si Negative Electrodes	LBNL
3	ES224	Fundamental Studies of Lithium-Sulfur Cell Chemistry	LBNL
3	ES225	Design and Synthesis of Advanced High-Energy Cathode Materials	LBNL
3	ES232	High Energy Density Electrodes via Modifications to the Inactive Components and Processing Conditions	LBNL
3	ES234	Electrode Materials Design and Failure Prediction	LBNL

continued

TABLE 3-18 Continued

EST Subactivity	Project ID	Project Name	Organization
3	ES233	Efficient Rechargeable Li/O_2 Batteries Utilizing Stable Inorganic Molten Salt Electrolytes	Liox
3	ES071	Design and Scalable Assembly of High-Density, Low-Tortuosity Electrodes	Massachusetts Institute of Technology
3	ES106	Studies on High-Capacity Cathodes for Advanced Lithium-Ion	ORNL
3	ES273	Composite Electrolyte to Stabilize Metallic Lithium Anodes	ORNL
3	ES276	Mechanical Properties at Protected Lithium Interface	ORNL
3	ES056	Development of High-Energy Cathode Materials	PNNL
3	ES144	Development of Silicon-Based High-Capacity Anodes	PNNL
3	ES226	Microscopy Investigation on the Fading Mechanism of Electrode Materials	PNNL
3	ES275	Lithium Dendrite Prevention for Lithium-Ion Batteries	PNNL
3	ES282	Development of High-Energy Lithium-Sulfur Batteries	PNNL
3	ES230	Design of Sulfur Cathodes for High-Energy Lithium-Sulfur Batteries	Stanford University
3	ES272	Pre-Lithiation of Battery Electrodes	Stanford University
3	ES274	Nanoscale Interfacial Engineering for Stable Lithium Metal Anodes	Stanford University
3	ES214	First Principles Modeling of SEI Formation on Bare and Surface/Additive Modified Silicon Anodes	Texas A&M University
3	ES283	Addressing Internal "Shuttle" Effect: Electrolyte Design and Cathode Morphology Evolution in Li-S Batteries	Texas A&M University
3	ES215	Analysis of Film Formation Chemistry on Silicon Anodes by Advanced In Situ and Operando Vibrational Spectroscopy	University of California, Berkeley
3	ES216	Optimization of Ion Transport in High-Energy Composite Cathodes	University of California, San Diego

TABLE 3-18 Continued

EST Subactivity	Project ID	Project Name	Organization
3	ES055	NMR and Pulse Field Gradient Studies of SEI and Electrode Structure	University of Cambridge
3	ES278	Overcoming Interfacial Impedance in Solid-State Batteries	University of Maryland
3	ES277	Solid Electrolytes for Solid-State and Lithium-Sulfur Batteries	University of Michigan
3	ES279	New Lamination and Doping Concepts for Enhanced Lithium-Sulfur Battery Performance	University of Pittsburgh
3	ES284	Statically and Dynamically Stable Lithium-Sulfur Batteries	University of Texas at Austin
3	ES285	Mechanistic Investigation for the Rechargeable Li-Sulfur Batteries	University of Wisconsin, Madison
4	ES201	Electrochemical Performance Testing	ANL
4	ES228	BatPaC Model Development	ANL
4	ES296	Development and Validation of a Simulation Tool to Predict the Combined Structural, Electrical, Electrochemical, and Thermal Responses of Automotive Batteries	Ford
4	ES202	INL Electrochemical Performance Testing	Idaho National Laboratory
4	ES204	Battery Thermal Characterization	NREL
4	ES294	Computer Aided Battery Engineering Consortium	NREL
4	ES295	Consortium for Advanced Battery Simulation (CABS)	ORNL
4	ES203	Battery Safety Testing	SNL
5	ES245	Low-Cost, Structurally Advanced Novel Electrode and Cell Manufacturing	24M Technologies
5	ES250	A Commercially Scalable Process for Silicon Anode Prelithiation	Amprius
5	ES167	Process Development and Scale-Up of Advanced Active Battery Materials	ANL
5	ES168	Process Development and Scale-Up of Critical Battery Materials	ANL
5	ES244	Low-Cost, High-Capacity, Non-Intercalation Chemistry Automotive Cells	Georgia Institute of Technology

continued

TABLE 3-18 Continued

EST Subactivity	Project ID	Project Name	Organization
5	ES268	Low-Cost Manufacturing of Advanced Silicon-Based Anode Materials	Group14
5	ES246	Advanced Drying Process for Lower Manufacturing Cost of Electrodes	Lambda Technologies
5	ES243	Dramatically Improve the Safety Performance of Li-Ion Battery Separators and Reduce the Manufacturing Cost Using UV Curing and High-Precision Coating Technologies	Miltec UV International
5	ES164	Thick Low-Cost, High-Power Lithium-Ion Electrodes via Aqueous Processing	ORNL
5	ES165	Performance Effects of Electrode Coating Defects and IR Thermography NDE for High-Energy Lithium-Ion Batteries	ORNL
5	ES207	Towards Solventless Processing of Thick Electron-Beam (EB) Cured LIB Cathodes	ORNL
5	ES266	Co-Extrusion (CoEx) for Cost Reduction of Advanced High-Energy-and-Power Battery Electrode Manufacturing	PARC
5	ES242	A Disruptive Concept for a Whole Family of New Battery Systems	Parthian Energy
5	ES263	Electrodeposition for Low-Cost, Water-Based Electrode Manufacturing	PPG
5	ES269	An Integrated Flame Spray Process for Low-Cost Production of Battery Materials	University of Missouri
5	ES264	Li-Ion Battery Anodes from Electrospun Nanoparticle/Conducting Polymer Nanofibers	Vanderbilt

NOTE: Acronyms defined in Appendix D.
SOURCE: Project numbers are from the Department of Energy's 2016 Annual Merit Review, see Cooper (2016a) and DOE (2016f).

for energy storage technologies and related activities is $103 million, up from $83 million in FY 2015 and $76 million in FY 2014. The 2016 budget is allocated as follows: battery technology development (32 percent), applied battery research (21 percent), battery materials research (17 percent), and solicitations (30 percent) (Howell and Snyder, 2016).

Current Status Versus Goals

In 2012 the EESTT, which consists primarily of members from DOE and USCAR OEMs, and includes investigators from the national laboratories, the battery industry, and universities, established a new set of goals and targets for electrochemical energy storage technologies. The *Electrochemical Energy Storage Technical Team Roadmap* was published in June 2013 and provides a summary of the targets for various plug-in electric vehicle applications and their approach of attaining those goals (U.S. DRIVE, 2013e). A detailed list of the targets can be found on the United States Consortium for Automotive Research (USCAR) site, including goals for PHEVs, BEVs, 12V start-stop battery systems, high-power low-energy storage systems, ultracapacitors, battery separators, and electrolytes (USCAR, 2014). These goals, targets, and guidelines are not only used for proposal solicitations but can also play an important role in providing directions to investigators for future research (see Recommendation 3-23).

Since the NRC's fourth review of the U.S. DRIVE Partnership, the focus of electrochemical energy storage R&D has shifted from high-power batteries for HEVs to higher energy systems for PHEVs and BEVs. This is an appropriate shift, since there are almost 4 million HEVs of over 50 models in the United States. Also, many of the technical advancements realized for high-energy systems can be optimized for high-power systems required for HEV applications.

PHEV40 Battery Meets Cost Target but Falls Short of Energy Density Goals

Significant progress has been made in the cost reduction of the PHEV40 battery and the 2020 price target of $3,400 was achieved (see Table 3-19).[15] This cost projection is derived from ANL's Battery Production and Cost model (BatPaC) using material costs and cell and pack designs input. The cost projection is based on a production volume of at least 100,000 batteries per year. This cost represents more than a 50 percent reduction in system production price of $6,850 in 2012 (Howell and Elder, 2012). The committee has determined that these costs are not totally out of line with values available in the open literature.

The batteries also meet most of the performance targets; they far exceed the power requirements, meet the cycle life, and are projected to meet the calendar life, but are deficient in the 2020 energy density targets by about 40 percent.

BEV Battery Falls Short of Cost and Most Performance Targets

A major U.S. DRIVE research target has been to reduce the production cost of a BEV battery to $125/kWh by 2020. The current cost projection based on the BatPaC model is estimated to be less than $268/kWh. This represents a significant

[15] Note that PHEV40 designates a vehicle that can drive in all-electric mode for 40 miles before the battery is depleted and the vehicle has to be powered by its internal combustion engine.

TABLE 3-19 Summary of Plug-in Hybrid Electric Vehicle (PHEV) Performance Status

	PHEV40 (40 mile AER)	
Energy Storage Targets	Target (2020)	Status (2016)
Discharge pulse power: 10 sec (kW)	38	~175
Regen pulse power: 10 sec (kW)	25	~115
Available energy (kWh)	11.6	11.6
Calendar life (year)	10+	10
Cycle life (deep cycles)	5,000	5,000
Maximum system weight (kg)	120	~219
Maximum system volume (l)	80	~132
System production price at 100,000 units/year	$3,400	~$3,352

SOURCE: Howell and Snyder (2016).

reduction (55 percent) from the 2012 estimate of greater than $600/kWh but is still twice as much as the 2020 target.

There have been improvements in most performance targets but more needs to be done to meet all of the 2020 goals. The energy density of the lithium-ion battery has improved and the current ranges for weight and volume overlap with the target ranges. The current system weight of 200-280 kg is close to the target range of 160-240 kg and represents a significant reduction from 2012 value of 500-750 kg. Similarly the battery volume of 90-140 liters is close to the target value of 80-120 liters and represents more than 50 percent reduction from the 2012 battery volume (see Table 3-20). Considerable improvements in cycle and calendar life of the lithium-ion battery have to be realized for the BEV application. Also, the operating temperature range is significantly less than the target, particularly at low temperatures.

Electric vehicles on the road today use first-generation lithium-ion batteries consisting of a graphite anode, layered oxide cathode, and blended carbonate-based electrolyte. The first-generation lithium-ion battery has an energy density of up to 150 Wh/kg. The DOE is working on the second generation of lithium-ion technology to increase the energy density of lithium-ion batteries. Extensive R&D is being conducted on new high-voltage (5 V versus 4 V) high-capacity (~300 mAh/g versus 140 mAh/g) cathode materials, higher-voltage electrolytes, and high-capacity silicon or tin-based graphite or intermetallic alloy anodes. These high-energy electrode materials promise to increase significantly the energy density and lower battery cost by reducing the amount of material used and the number of cells needed for the entire battery pack. This is illustrated in Figure 3-17,

TABLE 3-20 Summary of Battery Electric Vehicle (BEV) Battery Performance Status

Energy Storage Goals	Target AEV (2020)	Current (2016)
Equivalent electric range (miles)	200-300	✓
Discharge pulse power: 10 sec (kW)	80-120	✓
Regenerative pulse power: 10 sec (kW)	40	✓
Available energy (kWh)	40-60	✓
Recharge rate (kW)	1.20	✓
Calendar life (years)	10+	TBD
Cycle life (cycles)	1,000 deep cycles	500-600
Operating temperature range (°C)	–40 to 60	0-40
System weight (kg)	160-240	200-280
System volume (liters)	80-120	90-140
Production cost at 100,000 units/year	$125/kWh	<$268

NOTE: AEV is all-electric vehicle, the same as a BEV. TBD, to be determined.
SOURCE: Howell and Snyder (2016).

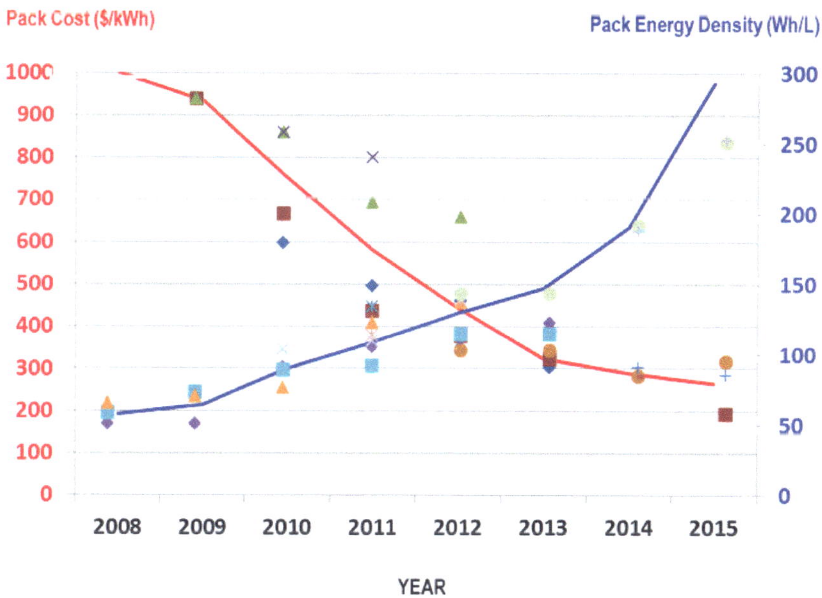

FIGURE 3-17 Energy density increase and resultant cost reduction for the plug-in hybrid electric vehicle 40 (PHEV40) lithium-ion battery.
SOURCE: Howell and Snyder (2016).

where the energy density is calculated from prototype cells and modules, and the battery cost is projected by the BatPaC model for the relevant electrode materials and projected battery design.

The results and estimates shown in Figure 3-17 are very encouraging; however, much more work is required to optimize all performance and lifetime requirements with reduced cost and abuse resistance in a single second-generation electrochemical couple.

Assessment of Progress and Key Achievements

The electrochemical energy storage program managed by the VTO with active collaboration with the USABC is a comprehensive program ranging from exploratory materials research to battery system development. It attempts to cover all aspects of energy storage development from materials, design, analysis, testing, manufacturing, and implementation for all electric drive applications such as HEVs, PHEVs, BEVs, and HFCVs. Significant gains have been made in cost and performance versus goals, which have been discussed in the "Current Status versus Goals" section and will not be repeated here. In recent years the focus of the effort has been appropriately directed toward high-energy systems particularly for PHEV and BEV applications. Furthermore, emphasis was given to high-energy materials, such as high-voltage cathodes and high-capacity anodes other than graphite, thus increasing the energy density of the system. For example, today's anodes have achieved 600 mAh/g vs. 300 mAh/g in 2012 and cathode capacities are 200 mAh/g vs. 120 mAh/g in 2012 (Howell and Snyder, 2016). This has allowed the use of less material and fewer modules for the required energy for each application and reduced cost. This has led to progress toward the development of the second generation of a lithium-ion battery with a significant projected cost reduction and represents a key achievement of this program. Additionally, there is slow but steady progress in the performance parameters for the lithium-ion battery.

There has been a conscious effort to investigate low-cost materials and processes as exemplified by the formation of Argonne's Materials Engineering Research Facility in support of the applied battery research subactivity. Significant progress has been realized in evaluating and testing new electrode materials through cooperative efforts of ANL and chemical producers, such as STREM and Aldrich, that can make reproducible materials in large quantities. Process improvements, such as reduction of solvent use in cathode coating by Johnson Controls and development of ultraviolet (UV) curable binders by Miltec, which will have a significant impact on the cost of the lithium-ion (Li-ion) battery, represent important progress by the program. The DOE/USABC contracts with Celgard and Enteck reduced Li-ion separator cost from $3/m^2 to approximately $1.20/m^2 and Amprius developed a full cell using a Si-nanowire anode and a nickel-manganese-cobalt cathode that delivers 306 Wh/kg with over 500 cycles (Howell and Snyder, 2016).

The VTO has funded the development of several modeling tools for use by the battery community. ANL developed the lithium-ion BatPaC for electric vehicles that is available to the public. BatPaC can predict the energy density and cost of an integrated battery pack as new battery materials are developed and newer design concepts are considered. The model predicts the impact of technology development at the component and process level on battery pack level performance and cost. It has increased confidence in the research direction of advanced lithium-ion technology to meet future goals.

NREL has introduced the Computer-Aided Engineering for Electric-Drive Vehicle Batteries (CAEBAT) activity to improve electric vehicle batteries through the development and validation of multiscale, multiphysics modeling tools. These have been used by more than 50 end users to simulate components, cells, and battery packs—for example, GM has used the simulation tool for battery pack thermal management and to predict optimum cell capacity for electric performance, cooling requirements, life, safety, and cost (Howell and Snyder, 2016).

A testament to the success of the DOE supported energy storage technology program is the increasing use of lithium-ion batteries in HEVs, PHEVs, and BEVs (DOE, 2015a). BMW, Mercedes, and Land Rover have introduced an HEV using lithium-ion batteries produced by Johnson Controls that were partially developed by the VTO-sponsored projects. LG Chem lithium-ion batteries that use the VTO-sponsored battery technology are used in the GM Chevrolet Volt and Cadillac ELR PHEV (Howell, 2016). LG Chem also supplies lithium-ion batteries for the GM Chevrolet Bolt and Ford Focus BEV.

The DOE is funding a number of projects to develop the next generation of electrolytes, which will require stability at the higher voltages anticipated by higher-voltage cathodes. A multidisciplinary high-energy/high-voltage cathode project involving six labs has been initiated and a silicon anode "deep dive" project consisting of five DOE labs has been initiated with a target capacity of 1,000 mAh/g and 1,000 EV cycles (Howell and Snyder, 2016). In addition, this approach may enable higher power densities thus reducing recharge times. The VTO also has a portfolio of projects to develop higher energy density batteries beyond lithium ion. The most prominent examples are the lithium-oxygen and lithium-sulfur chemical couples. These systems promise the potential for higher energy density but require major technical breakthroughs for success, particularly the use of lithium metal and its related safety and interfaces issues. These projects may benefit from the use of the various modeling technologies to guide the most promising research approaches.

A program has also been initiated to investigate solid-state electrolytes for lithium batteries. In recent years, high conductivities ($>10^{-4}$ S/cm for solid electrolytes at room temperature) have been achieved in inorganic solid electrolytes and are being considered for large-capacity all-solid-state batteries as alternatives to liquid or polymer lithium batteries. Solid electrolytes, in addition to superior safety characteristics, may also result in increased energy density since they present a wider voltage window for higher-voltage batteries and the possibility of bipolar construction to form battery packs and minimize the amount of inactive materials used. Higher solid-state electrolyte conductivity and lower electrolyte/electrode interface impedances will be needed to realize these advantages for both lithium metal-based and advanced lithium-ion batteries.

Significant Barriers and Issues That Need to Be Addressed

A major barrier to electric drive automobile market penetration is the high cost and the low energy density of batteries. The battery comprises the highest cost component of the electric propulsion system for HEVs, PHEVs, and BEVs. Although there has been a significant improvement in lithium-ion battery energy density that may be sufficient for HEVs and PHEVs, additional improvement is required for BEVs. Energy density and cost are highly correlated: an increase in energy density generally results in a significant cost reduction since higher-energy systems require less material, which reduces materials costs, and smaller packages, which reduce processing costs. Furthermore, systems with high energy can be optimized for high power, which is required for HEV and PHEV applications.

In addition to cost and performance characteristics, safety remains a significant issue with lithium batteries. As noted in the NRC Phase 4 report (NRC, 2013), there has been great progress in recent years in overcoming safety issues through the use of intrinsically safer materials and the design for safety on a systems level. As newer higher energy density systems are developed, continued diligence is needed to test for abuse tolerance given the tendency for increased hazard with higher energy density systems. It is also important to be aware of safety issues arising from actual road use as witnessed by recent fires of the Tesla S caused by roadside debris (Larsson et al., 2014) or fires of the GM Volt after safety testing (Smith, 2012).

Response to Recommendations from the NRC Phase 4 Report

> ***NRC Phase 4 Recommendation 3-12.*** While continuing mainstream efforts to increase energy density and reduce the cost of high energy batteries for BEV and HEV applications, the U.S. DRIVE Partnership should intensify its development of high-power batteries and supercapacitors as such technology impacts all types of hybrid vehicles (HEVs, PHEVs, and HFCVs). It should also more closely integrate its efforts with other DOE offices and agencies to investigate new high energy electrochemical couples for BEV applications.
>
> ***Partnership Response.*** U.S. DRIVE places major emphasis on the development of high-power batteries and supercapacitors, and the U. S. Department of Energy's (DOE) activities align with this approach. Examples include the following:
>
> DOE's Vehicle Technologies Office (VTO) issued a funding opportunity announcement in March 2013 that identified insufficient power density as a barrier associated with electric drive vehicles and sought projects to develop batteries with high volumetric (1600 W/L) and high gravimetric (800 W/kg) power density.
>
> VTO, with the United States Advanced Battery Consortium (USABC), is currently funding a project with Maxwell Technologies to develop a hybrid ultracapacitor sys-

tem for the Low Energy-Energy Storage System (LEESS) for (HEVs) and completed a LEESS technology assessment program with Actacell. The funders also initiated a project with Saft America to develop a high-power, 12V start/stop battery.

VTO supports applied research programs to improve the high power lithium titanate anode material and develop electrode/electrolyte materials that will enable an ultracapacitor to meet the USABC power goals.

DOE national laboratories continue to benchmark batteries to assess their capacity to meet HEV, LEESS, or 12V start/stop requirements, including products from Axion, A123 Systems, Johnson Controls, Maxwell, and Hydro Quebec.

DOE coordinates its work on these topics both internally across offices and externally with other federal agencies. DOE's Integrated Battery Technical Team, which includes participants from VTO, ARPA-E (Advanced Research Projects Agency-Energy), and the Office of Science, ensures that research and development (R&D) goals and activities are well coordinated across the Department. Interagency efforts include the Chemical Working Group of the Interagency Advanced Power Group (IAPG), active participation in program reviews and technical meetings other federal agencies sponsor, and similarly, other agency expert participation in the contract and program reviews of DOE-sponsored efforts. DOE also coordinates with the U.S. Department of Transportation/National Highway Traffic Safety Administration (DOT/NHTSA), the Environmental Protection Agency, and the United Nations Working Group on Battery Shipment Requirements. DOE also cosponsors battery R&D with other agencies when appropriate. For example, VTO's March 2013 funding opportunity includes plans for a U.S. Army contribution of $3.5 million in co-funding for several areas with joint development opportunities.

Committee Assessment of Response to 3-12. The committee thanks the U.S. DRIVE Partnership for the comprehensiveness of their response and is pleased with the ongoing effort.

NRC Phase 4 Recommendations 3-13 and S-6. The USABC targets for BEV batteries are more than 20 years old and should be revised, as also recommended in the NRC's Phase 3 review. U.S. DRIVE should also undertake a diligent effort to develop a consistent set of technical targets across the key electric drive vehicle applications.

Partnership Response. The Committee is correct in noting that the USABC targets for BEV batteries are more than 20 years old. However, over the past 20 years, the USABC has reviewed the analyses and assumptions that produced those requirements and has found them to be still valid. Thus, it should be noted that the requirements went unchanged because they remained relevant and valid. Recently, the U.S. DRIVE Partnership took a number of steps to revise the energy storage targets for electric vehicles and have included these targets in the new Electrochemical Energy Storage Technical Team (EESTT) Roadmap. As the committee notes, the existing requirements are in a form that is inconsistent with more recently published requirements for plug-in hybrid electric vehicles (PHEVs) and hybrid electric vehicles (HEVs). For example, the BEV battery pack targets are related in terms of energy density, as compared to PHEV targets that provide total pack energy requirements. Over the last few years, there has been growing interest in the development and commercialization of specific BEV vehicle platforms with long and intermediate range (such as a mid-size sedan with a 100 or 300 mile range). Thus, the Partnership took a number of steps to revise the energy storage targets for electric vehicles to reflect specific pack level performance targets. Argonne National Laboratory and the National Renewable Energy Laboratory developed a detailed analysis of energy storage requirements through simulation/modeling, which will be documented in the U.S. Department of Energy 2012 Annual Progress Report on Energy Storage R&D. The USABC is reviewing these analyses and formal targets are expected to be approved in 2013. U.S. DRIVE has adopted a consistent set of energy storage targets for fuel cell/battery hybrid vehicles and internal combustion engine/battery hybrid vehicles. The energy storage system must accept the same regenerative braking energy and provide the same power assist (depending on the level of hybridization), regardless of whether it is hybridized with an internal combustion engine or a fuel cell. This was the conclusion of both the Electrochemical Energy Storage and the Fuel Cell Technical Teams after extensive analysis by a joint working group of both teams.

Committee Assessment of Response to 3-13 and S-6. After several comments and recommendations, DOE has finally revisited the USABC targets and published a revised roadmap for the various electric vehicle types. This is a positive step in the right direction; however, there are yet several inconsistencies in the targets and goals in various publications including the following:

- DOE has established a $125/kWh cost target for the BEV battery. No precise date for this goal is mentioned in the EESTT roadmap (U.S. DRIVE, 2013c). Howell in the April 2016 presentation to the committee mentioned that the goal is for 2020. The EV-Everywhere program (DOE, 2014b) and the annual progress reports for 2014 and 2015 present this goal for 2022 (DOE, 2014a; DOE, 2015a).
- The 10-s discharge power is listed as 38 kW in the USABC PHEV40 goals (USCAR, 2014) and in David Howell's presentation on April 2016, but a value of 50 kW is quoted in the EESTT roadmap (U.S. DRIVE, 2013c).
- The recharge rate is listed as 1.4-2.8 kW in the EESTT roadmap (U.S. DRIVE, 2013c) versus 3.3 kW (240 V/16 A) in the USABC PHEV40 goal (USCAR, 2014).

Such inconsistency in values in the various publications can cause confusion over what the goals and targets are. It would be good to have a convenient link to regularly updated goals and targets in the VTO website that all can reference as the real values. In addition, dates should be associated with these targets and goals and, if left unchanged, a last review date should be provided. (See Recommendation 3-23.)

In 2012 the EESTT established new goals and targets for the 2020 time frame (2018 or 2022 depending on the application and the publication). As this report is issued in early 2017, 2020 is very close at hand. It may be time to revisit the goals and targets given the advances in battery technology and vehicle implementation in the last 4 years. The roadmap does mention in passing that there is an opportunity to double the energy density of lithium ion in the near term (2012 to 2017), and for the long term (2017-2027) "beyond lithium ion" battery chemistries may be required. It would be more helpful to provide performance goals and targets for rolling decades 2020, 2030, and 2040 such that investigators can look for the appropriate chemistries to meet the necessary performance goals.

Appropriate Federal Role

The long-term R&D aimed at fundamental discoveries of new materials and chemistries beyond lithium ion is absolutely appropriate for federal funding. This work is of national interest and will not only result in new discoveries but also produce the next generation of energy storage scientists, as the work is conducted at several universities. The focus by the applied battery research subactivity on optimizing the next generation of electrochemical couples and manufacturing technologies is an appropriate use of the resources at the national laboratories. The battery technology development subactivity coordinated with USABC is of a more near-term nature, involves battery module and system development, and is conducted with 50 percent cost share by the vendors. The cost share model ensures that work is conducted on relevant technologies. The committee does wonder, however, if the increasing effort on benchmarking and testing of battery products is an appropriate role for DOE and if any necessary information required for precompetitive research cannot be obtained with some agreement with the USABC OEMs.

Findings and Recommendations

Finding 3-35. The Vehicle Technologies Office (VTO) maintains several modeling tools, such as the Battery Production and Cost model, the Computer-Aided Engineering for Electric-Drive Vehicle Batteries, and others, to accelerate the development of improved batteries for electric vehicles. These tools can be used to predict energy density and battery pack costs for new battery materials. The VTO is also investigating lithium-oxygen and lithium-sulfur batteries and looking for higher energy density batteries beyond lithium ion.

Recommendation 3-22. DOE should apply these modeling tools to the advanced lithium-based chemical couples it is investigating to determine which systems should be continued and which should be stopped and the appropriate direction to pursue to provide the best possibility of meeting their performance, cost, and safety requirements. The modeling tools should also be used to study other non-lithium systems that may be appropriate for the various plug-in electric vehicle applications and DOE should only look at the most appropriate systems.

Finding 3-36. After a 20-year gap, a new set of energy storage goals and targets for the various electric vehicles was established in 2012. These targets are to be realized in the year 2020. This was a very positive step; however, all of the goals and targets are not in one place and there are several inconsistencies in the target values in various publications. It may be time to revisit the goals and targets given the advances in battery technology and vehicle implementation in the last four years.

Recommendation 3-23. U.S. DRIVE should establish a single, authoritative website for energy storage targets and goals for the various electric vehicle applications that is prominently and easily accessible to all. The dates that targets and goals were set or reviewed without change should be provided. The site should provide a roadmap of energy storage needs for several (rolling) decades into the future for use by research organizations and investigators for various applications and differing time frames.

Finding 3-37. Recent advances in inorganic solid-state electrolytes provide an alternative way to increase safety, energy density, and cost of not only lithium-metal based batteries but also for advanced lithium-ion batteries.

Recommendation 3-24. The Department of Energy should increase the effort on solid-state electrolytes, in particular to look for newer materials with higher conductivity and address the high impedance at electrolyte/electrode interfaces for these materials.

ELECTRICITY AS AN ENERGY SOURCE AND THE GRID

The convenience, affordability, and environmental impacts of electric energy have become important considerations for the U.S. DRIVE Partnership. Most obviously, the environmental and energy security benefits from BEVs and PHEVs will increase in proportion to their use, commonly measured in electric vehicle miles traveled. And so, the availability and cost of recharging options weighs importantly in consumer decisions to purchase and use plug-in electric vehicles.

HFCVs also interact with the electric grid. These could serve as a backup electric supply with a typical automotive power train, about 70 kW, able to serve a small cluster of homes.[16] More importantly, the pathways for producing hydrogen for fuel-cell vehicles include the electrolysis of water. Electrolysis (also termed "water splitting") produces hydrogen gas by passing an electric current through water. The economics of producing hydrogen through electrolysis depend on both the efficiency of the process and the cost of the electricity. At the efficiency typical of current electrolyzers with ongoing improvements that can reasonably be anticipated, 50 to 65 kWh of electricity would be required to produce a kilogram of hydrogen. At the average U.S. electricity cost of around \$0.11/kWh, this ranges from \$5.50/kg H_2 to around \$7.00/kg H_2, far too costly to compete with other production technologies (see the section "Hydrogen Production, Delivery, and Distribution," earlier in Chapter 3).

However, the economics of hydrogen production from electricity also depend on the cost of the electric energy used, and with renewable energy, the averages

[16] This approach is most evident in Japan and is consistent with the heavy government subsidies for HFCVs and the enthusiasm of the Japanese automotive OEMs for them.

can be misleading. For wind, for example, energy production is strongest during the nighttime hours when the demand for electricity is lowest. As a consequence, the grid value of electric energy is also at its daily low. For example, an opportunity cost for this off-peak electricity around $0.02/kWh would yield an energy cost of approximately $1.00/kg to $1.40/kg of hydrogen—to which, of course, capital and O&M costs would have to be added.[17] All this suggests the potential of grid electricity to contribute to hydrogen production as well as to recharging for BEVs and PHEVs.

At the same time the electric grid is beginning a long-term transition away from large coal-fired power plants toward more use of natural gas[18] and greater reliance on cleaner but intermittent energy sources, chiefly solar and wind energy. Some of this new generation, especially from wind, will be utility scale, but solar energy could also become important at the edge of the grid with smaller units placed on residential or commercial rooftops. Accommodating this fuels transition will require major changes in the transmission and distribution portions of the grid, the "wires" infrastructures that connect the sources of energy with the customers for that energy. Energy storage must increase greatly to accommodate the intermittent energy sources. And for greatest efficiency, power must be tradable among grid participants who are alternately sources and users of electricity.

The pace and direction of this transition is largely being set at the state rather than the federal level because state regulatory commissions enjoy almost exclusive jurisdiction over the rates charged and services allowed for retail energy. The decisions of these regulators will strongly influence the cost, convenience, and environmental impact of electric energy used for advanced vehicles.

The U.S. Electric Sector: The Case of the Plug-in Electric Vehicle

The focus here is on plug-in electric vehicles because of their immediacy and prominence in the advanced vehicle marketplace. A National Research Council study (NRC, 2015a) noted the striking diversity of the electric vehicle refueling infrastructures in contrast with petroleum refueling. Recharging opportunities include the following:

- Home charging, typically Level 1 AC (120 V) or Level 2 AC (240 V);
- Workplace charging, Levels 1 or 2 AC and possibly higher;
- Over-the-road charging (intracity, intercity, or interstate) often requiring power levels as high as 120 kW per vehicle charger; and
- Wireless charging, which uses an inductive coupling to transfer energy but has yet to find widespread application.

[17] The use of off-peak electricity for fuel cell vehicles would, of course, have to compete with other uses such as pipeline or industrial gas, and with battery storage for on-peak grid use.

[18] In 2015, coal contributed 33 percent of U.S. electricity generation, natural gas another 33 percent, and petroleum about 1 percent (EIA, 2016b).

In considering this diversity of energy and power requirements, the NRC committee examined each component of the U S. electric system separately. The generation and transmission components were found quite adequate for current and anticipated levels of BEV and PHEV use. However, the distribution system could face challenges under some circumstances:

- Several plug-in electric vehicles charging on a single branch circuit designed for traditional home loads could overload that circuit.
- A lesser number of vehicles charging during times of peak demand could have the same effect.

To be sure, technical remedies for these difficulties are readily at hand, the solutions ranging from upgrading circuit capacity to incentives built into electric rates for off-peak vehicle charging. However, the economic incentives for these remedies will depend on the rulings of the state regulatory commissions.

Finally, the NRC study concluded that the electric utilities can find a legitimate business case for promoting electric vehicle charging within their service territories. But once again, the state public utility commissions will need to provide regulatory incentives that ensure that benefit accrues to both plug-in electric vehicle owners and the remainder of the utilities' customers.

Evolution of the U.S. Electric Grid: Implications for the U.S. DRIVE Partnership

The electric grid seems likely to evolve well beyond its present form as new paradigms for cost and service emerge. The pathway taken will prove consequential for both plug-in electric vehicles and for HFCVs, offering new opportunities and new challenges. For example, the term "transactive energy" (TE) has come into use as a general descriptor for a more flexible electric grid that enables real-time trading among intermittent sources and users of electric energy. The GridWise Architecture Council defines TE in most general terms as "a system of economic and control mechanisms that allows the dynamic balance of supply and demand across the entire electrical infrastructure using value as a key operational parameter" (Forfia et al., 2016).

From an operational perspective TE services would be delivered through an information technology platform that integrates suppliers and buyers of energy in a marketplace with extensive synergies shared among all participants and with the platform provider. This platform would enable customized, on-demand energy services and derive its synergies from the elimination of intermediaries and from the demand for new and innovative services (Masiello and Agüero, 2016).

These incipient changes could be important to the U.S. DRIVE Partnership since the vehicle (either all-electric or hydrogen fuel cell) could become a participant in some future TE system, at times a customer for energy services (especially battery electric vehicles) and at other times a provider of services (capacity, reliability, or demand response, for example). To the extent that their design can accommodate these additional requirements, the vehicles will be more highly valued by consumers and their deployment accelerated. But to the extent that providing grid services reduces the usefulness of the vehicle—for example by shortening battery life through more frequent charge-discharge cycles—the vehicle owners might find that the revenues from such services fail to offset the reduction in vehicle utility.

An organization called the GridWise Architecture Council, funded in part by the Department of Energy, has developed and continues to refine the overall paradigm for TE (GridWise Architecture Council, 2015). Two kinds of implementation strategies are being pursued to bring TE to realization in electric markets. The first, a top-down strategy, can be seen in *Reforming the Energy Vision* (REV), an initiative by the state of New York. The intent is to use policy and regulation to build the TE marketplace and thereby provide incentives for private investment in distributed energy resources (DER)[19] (Zibelman, 2016).

In contrast, the state of California has taken a "bottom up" approach because prior policy and regulatory actions have already led to ample DER. It seeks to build a TE marketplace to better integrate those DERs into wholesale markets and operations while preserving grid reliability (Masiello, 2016).

Other states offer variants on these archetypes. In Hawaii, generous incentives for DER have increased rooftop solar installations to the point that the photovoltaic (PV) output exceeds the daytime feeder loads. In response, the state utility commission has directed investor-owned utilities to propose grid-scale storage projects. In contrast, Massachusetts with limited DER installations requires smart metering and time-of-day pricing in planning its distribution system. But in every case, significant issues remain to be resolved before the promise of transactive energy can be realized. These include the following:

- The fundamental nature of the control system, which ranges from a centralized, system-wide optimization performed centrally to a decentralized, layered optimization structure (Kristov et al., 2016);
- The transition conditions which must maintain electric service while changing from one fundamental paradigm to another;
- Incentives for investment in long-lived assets in a highly dynamic system; and
- Grid security in a more open-architecture system.

[19] The term "distributed energy resources" most commonly includes renewable energy, chiefly wind and solar; battery energy storage systems; and demand response, electric loads that can be deferred in time, which creates a "virtual power plant."

Assessment of Progress

The grid interaction technical team (GITT) participants include DOE, USCAR, FCA, Ford, GM, DTE Energy, SCE, and EPRI. MISO and Eversource are associate members.[20] The mission of GITT, as described by the team, is to "support a transition scenario to large scale electrified vehicle charging with transformational technology, proof of concept and information dissemination" (Slezak and Gross, 2016). It entails a "collaborative effort to address the interests of U.S. DRIVE partners and other stakeholders to identify and reduce barriers to large scale introduction of grid connected vehicles" (Slezak and Gross, 2016).

The scope claimed by GITT is "the interaction between the electric power grid and light duty electric/plug-in hybrid vehicles through the charging infrastructure, and focuses on the following areas:

- Developing standards enabling interoperability;
- Consumer usability;
- Life cycle/total cost of ownership;
- Interface of the plug-in vehicle to the local power distribution network (physical);
- Electric distribution/modern grid interface (IT); and
- Use case scenario, grid scenarios planning, feasible for EV vs. grid vs. aggregator vs. workplace" (Slezak and Gross, 2016).

By design and as noted in the presentation to the committee by the GITT, this technical team does not have technology targets to measure its progress. The committee understands why this is appropriate for a team that is facilitating interactions among a variety of stakeholders (the facilitator's role) in a complex, interactive system. But on the other hand, the apparent absence of accountability, that is, the lack of technical targets, works against the ability of the facilitator, that is, the GITT, to learn the lessons of experience and measure its progress.

In its presentation to the committee, the team presented a strategy, a list of R&D tasks and projects, and a set of future plans. Its strategy covers a broad range of concerns (Slezak and Gross, 2016):

- Develop/verify technology to facilitate grid integration
  - —Consistent with DOE Grid Modernization Initiative
  - —Cybersecurity
  - —Communication and control technology for grid integration
  - —Submetering
- Identify and implement what the market needs to be successful

[20] Tesla was a member but withdrew from the Partnership in July 2016.

- Develop and use test fixtures to support SAE vehicle-grid interface standards development
  - Communication, interoperability, wireless charging, power quality
- Data collection and evaluation to fill knowledge void
  - User experience
- Leverage activities to support global cooperation, harmonized standards, and component compatibility
  - U.S.-EU and U.S.-China cooperative agreements
  - Joint activities—pilot projects to facilitate harmonization of standards
  - Standard laboratory test procedures and protocols
- Promote industrial collaboration
  - Expand reach and relationship to include newer groups to include larger system that EVs are tied into
  - Alignment of EVs, grid, and buildings

Its projects address a variety of related areas: cost reduction, convenient charging locations and workplace charging, smart charging, electric vehicle supply equipment compatibility and interoperability, and plug-in electric vehicle and grid integration. The team noted it was focused on near-term implementation with assumed long-term impact. The committee is concerned that the variety of efforts lacks apparent coordination, coherence, and accountability.

Consider this example of a goal formally (but rarely) set: the declaration in 2013 that in its roadmap under the topic Smart Charging/Smart Grid Interface, "EV-smart grid communication and smart energy management will be demonstrated using hardware-in-the-loop techniques with a simulated grid in 2013; this will potentially be followed by a collaborative grid integration activity in 2014-2015" (U.S. DRIVE, 2013e). Two aspects of this statement raise concern: (1) the statement appears in passive voice with no attribution of responsibility, and (2) the committee can find no follow-up indicating the outcome of this expectation. And without either some notion of responsibility or some follow-up, it is difficult to see how the lessons of experience can be learned.

Further, the committee is unable to discern a logical connection among the "Major Challenges and Barriers" that the GITT sees ahead of it, the "Strategy" that the GITT sets for itself, the "Goals and Tasks," and the "R&D Tasks/Projects" through which the GITT seeks to implement.[21] It would seem that these should be related in some logical way if the facilitation is to be effective.

To be sure, the committee is fully aware of the difficulties of a "facilitator's" role. Nevertheless, the prevalent ambiguity leads to the appearance (and perhaps the reality) that these activities lack strategic clarity and intent.

[21] These titles refer to briefing charts in the technical team's presentation to the committee (Slezak and Gross, 2016).

Response to NRC Phase 4 Recommendation

NRC Phase 4 Recommendation 2-9. The grid interaction technical team should make a special effort to work with utility regulatory commissions throughout the United States to (1) help identify the best practices in rate regulation that could advance the deployment of plug-in vehicles if widely used, and (2) communicate the advantages from these best practices accruing to the public and to state and local officials.

Partnership Response. It is important to note that U.S. DRIVE focuses on precompetitive R&D, and as such, policy- or education-related activities are considered out of scope. However, outside of the Partnership activity, Grid Interaction Technical Team (GITT) members will continue earlier and existing efforts and work with the utility regulatory commissions to identify "best practices in rate regulation" and communicate these best practices to industry stakeholders. Continuing these endeavors will expand on efforts to develop streamlined electric vehicle infrastructure permitting and inspection processes. GITT members, individually and as part of a united effort with the Edison Electric Institute, have been engaging with the National Association of Regulatory Utility Commissioners for this purpose and will support efforts to identify and share best practices with the utility commissions through various outreach channels, including those involving the U.S. Department of Energy.

Committee Assessment of Response to 2-9. The committee appreciates and agrees with the need for focus on precompetitive R&D. The committee would, however, suggest that insights into the directions that state regulators are advocating for the electric industry can inform such R&D. This incipient change is not being pursued uniformly across the country. A few state regulatory commissions, notably California and New York, are most active in initiating the integration of renewable energy into the electric grid. This integration will increase the environmental benefits from the use of plug-in electric and hydrogen-fueled vehicles. But it will also influence the ability of the grid to support the deployment of these vehicles. Thus the committee continues to recommend the GITT establish close liaison with these commissions and systematically disseminate the lessons learned to the U.S. DRIVE partners and constituents.

Findings and Recommendations

Finding 3-38. The electric grid is beginning a period of disruptive change brought on by (1) technological opportunity, especially in microelectronics, deep learning, and robotics; (2) the global need to reduce the carbon emissions from electricity production; (3) marketplace demand for new and more efficient energy services; and (4) threats from outsiders to the security of a more automated grid linked to external devices. As a consequence, the electric grid will continue to evolve in ways that are not predictable to either the incumbents or the disruptors. State regulatory authorities will shape the pace and direction of this transition to a greater extent than the federal government.

Recommendation 3-25. The U.S. DRIVE partners should closely monitor the evolution of the electric grid to understand how (or whether) vehicle design can enable effective participation in the emerging electric marketplace in a way that increases the market share of nonpetroleum vehicles such as hydrogen fuel cell vehicles and (possibly) battery electric vehicles.

Recommendation 3-26. The grid interaction technical team should focus on and learn from the activities of the state regulatory commissions who have emerged as the leaders and promoters of the electric system transition.

Recommendation 3-27. The grid interaction technical team should seek to learn from the experience of the participants in grid modernization and vehicle-to-grid linkages, formally document that learning, and make those documents available to all constituencies.

Finding 3-39. The Partnership needs an honest facilitator for communicating needs, constraints, and aspirations among the complex set of stakeholders that link the electric grid with vehicles. Strategic clarity is essential for any participant in such a complex system, but especially so for the facilitator. The tasks of the facilitator include learning the lessons of experience and communicating those to the participants. This cannot be achieved without some understanding of accountability for promised actions.

Recommendations 3-28. The grid interaction technical team should explicitly relate its own activities to the various elements of its program ("Major Challenges and Barriers," "Strategy," "Goals and Tasks," and "R&D Tasks/Projects"), and make explicit its own activities and goals in a way that is traceable and accountable.

LIGHTWEIGHTING MATERIALS

Background

A major approach for improving vehicle efficiency, and thus fuel economy, is reducing the mass (DOE, 2010; EPA, 2014; Joost, 2012). A midsize family car weighs about 1,450 kg. A weight reduction of approximately 150 kg, or 10 percent, would achieve a 3 to 6 percent improvement in fuel economy. As noted in the remainder of this paragraph (Taub and Luo, 2015), and as seen in Figure 3-18, following the global oil crisis of the 1970s, the weight of automobiles decreased consistently for about one decade. The weight reduction was achieved by a combination of a shift from body-on-frame to body-frame-integral architectures, improved modeling tools, and the introduction of lightweighting materials. This era was followed by a period of stable oil prices and, in the

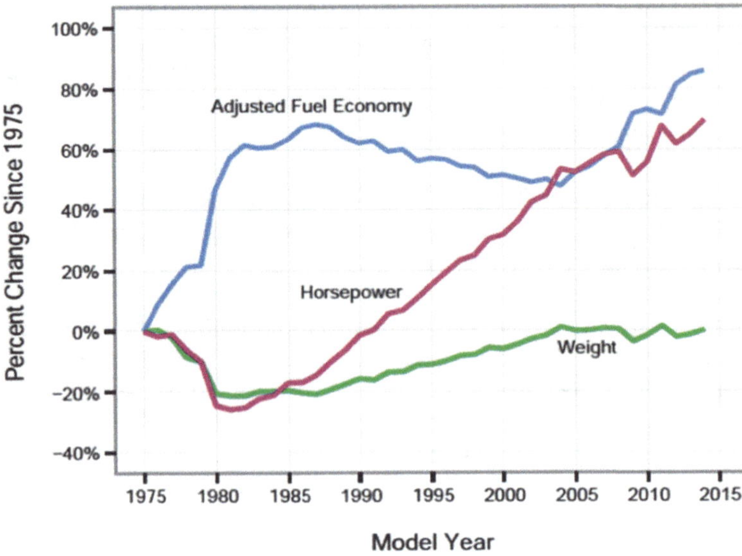

FIGURE 3-18 Changes in adjusted fuel economy, horsepower, and weight in the U.S. automotive market for model years 1975-2014. SOURCE: EPA (2014).

North American market, a shift to larger and heavier vehicles. Since the 1990s, engineering improvements in vehicle structural efficiency have continued, but the improvements have been offset by a consumer shift to sport utility vehicles, increased safety features, and other consumer-driven content, such as convenience features and infotainment systems. More recently, higher fuel economy standards are being adopted worldwide, and the newest vehicle models (e.g., the Cadillac ATS; Ford 150) are exhibiting weight reductions of 5 to 10 percent or more (Taub and Luo, 2015; Brooke et al., 2016). These weight reductions are enabled by the introduction of advanced high-strength steels, aluminum, magnesium, and polymer composites.

Status of U.S. DRIVE Materials Goals and Targets

As reported in the NRC Phase 4 review, the U.S. DRIVE materials technical team (MTT), like the FreedomCAR and Fuel Partnership before it, adopted a stretch goal of 50 percent reduction in vehicle weight versus 2002 comparable vehicles (NRC, 2013). Reducing vehicle weight is critical to improving fuel economy and doing so at the least incremental cost, while challenging, is worthy of pursuit, since achieving the 50 percent goal would result in up to a 35 percent fuel economy improvement although the literature does not support what fuel economy would be achieved with a 50 percent weight reduction.

The committee feels that the scope defined by the MTT is realistic, well stated, and both necessary and sufficient to meet this target:

- Materials research including but not limited to light metals (aluminum and magnesium), advanced high-strength steels, polymer composites, and mixed material systems to enable the U.S. DRIVE Partnership to reach its goals.
- Development of advanced manufacturing processes including stamping, casting, hydroforming, extruding, and injection molding to facilitate the widespread use of lightweight materials.
- Enabling technology development including optimal designs, advanced joining and assembly, corrosion mitigation, crash energy management, predictive and computational tools.
- Critical review of white paper studies on weight savings, vehicle performance and cost estimates.

MTT technical strategy (bold provided by MTT):

- **Focus on stretch but realistic goals and objectives** to develop lightweight, high-performance, cost-effective structural materials for vehicle lightweighting.
- **Engage** the steel, aluminum, magnesium, carbon fiber, polymer composite, and plastic resin suppliers.
- Work closely with supply base to **develop infrastructure of advanced manufacturing enablers** in (forming, casting, molding, joining, and assembly) for advanced light material systems for automotive applications.
- Deliver **computational tools and methods**, reducing the time and cost in developing and validating new materials, processes, and manufacturing technologies.
- **Engage national laboratories, academic institutions, and industrial research laboratories partners** in the integration of constitutive models ranging from fundamental alloy development to advanced manufacturing processes.

The committee strongly supports the strategy of working closely at an early stage with the supply base. This will be critical to ensure that the technologies developed under the U.S. DRIVE Partnership can be transitioned to the suppliers for commercial implementation. The committee also encourages expanding the development of computational tools as part of the larger Integrated Computational Materials Engineering (ICME) initiative.

At the same time, the first bullet of the strategy states that the MTT should be focused on "stretch but realistic goals." The MTT target of achieving a 50 percent weight reduction also states that this should be done *with equal affordability* (emphasis added) (NRC, 2013). Previous NRC committees (NRC, 2008; 2010; 2013) found the equal affordability goal to be unrealistic. During this review, the MTT reported a revision to the target by setting the comparator as a 2015 vehicle. The committee feels that this makes the affordability target even more unrealistic. The major concern is that there can be consequences on the work scope and at the project level if the targets are not appropriate.

As noted in Taub and Luo (2015), who reference Verbrugge et al. (2009) as well, the options required for weight reduction need to be compared with other fuel economy improvement solutions including increased power train efficiency, vehicle electrification, decreased tire rolling resistance, and improved aerodynamics. This has resulted in a general industry short-term target for the cost of weight reduction of ≤$4.5/kg saved ($2/lb saved) for weight reduction to be competitive with the other fuel-economy actions.[22] It appears to be more appropriate to set a long-term target for weight reduction for the Partnership, in a similar manner. That would involve comparing the relative costs for fuel economy improvement.

For this review the MTT reported adoption of a midterm target for 2020 of an 18 percent weight glider weight reduction to be achieved at <$11/kg saved (<$5/pound saved) while maintaining equal vehicle level performance (crash; noise, vibration, and harshness [NVH]; durability; reliability; and recyclability) (see Figure 3-19). The committee is pleased to see this more realistic goal, which allows the Partnership's activities to remain focused on enabling advanced high-strength lightweight materials and reducing their cost, even if the ultimate stretch goal is unrealistic.

Progress and Key Achievements

The technical accomplishments in materials of the Partnership for the years 2013-2015 are available on the USCAR website.[23] The MTT highlighted five major areas of achievement in its presentation to the committee on June 22, 2016 (Zaluzec and Joost, 2016). These were as follows:

1. *Multi-Material Lightweight Vehicle (MMLV) Project*
   - Objective: Engineer, build and test a lightweight multimaterial "concept vehicle" using commercially available lightweight materials and manufacturing technologies, capable of making 250,000 vehicles per year.
     - The Mach I prototype design achieved approximately 15 percent weight savings by incorporating modifications to the body, closures, interior, chassis, and engine components.
     - The estimated variable cost of the weight save is $5.15/pound. This result is an important demonstration toward achieving the midterm 2020 target.

2. *Virtual Materials Modeling*
   - Objective: Development and validation of carbon fiber composite material models for crash simulation.
     - The PAM-CRASH model was validated for a front bumper and crush can system. This result is a critical step toward implementation of carbon-fiber composites in primary crash structural applications and on the MTT path to achieve the mid-term 2020 target.

[22] The cost metrics reflect both amortized capital investment and material cost.

[23] The website for USCAR is http://www.uscar.org/guest/partnership/1/us-drive, accessed September 19, 2016.

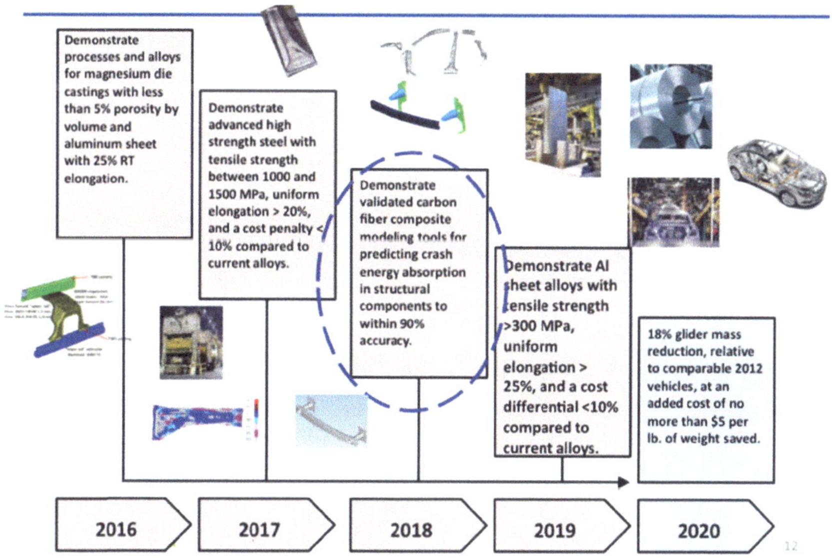

FIGURE 3-19 Significant materials technical team (MTT) target setting (2016-2020). SOURCE: Zaluzec and Joost (2016).

3. *Third-Generation Advanced High-Strength Steel (3GAHSS)*
   - Objective: Employ Integrated Computational Materials Engineering (ICME) to identify and assemble length-scale material models for predicting 3GAHSS constitutive behavior
     - —Progress was made in applying ICME modeling to a third-generation advanced high-strength steel that can predict constitutive behavior. This result is also on the MTT path to achieve the midterm 2020 target.

4. *Plasma-generated oxygen ionic species were shown to reduce oxidation phase of carbon fiber processing by over 60 percent.*
   - 4M Industrial Oxidation is working to commercialize this technology.

5. *Friction stir joining of tailor-welded blanks was increased from an original 1 meter per second up to 6 meters per second for production welding speed.*
   - The technology is being implemented at the TWB facility in Monroe, Michigan.

Significant Barriers and Issues That Need to Be Addressed

The MTT did a good job of capturing the challenges to reaching the targeted 50 percent weight reduction target. Key challenges are as follows:

- *Mass increases due to regulations (NHTSA, FMVSS, EPA, CARB) including hybridization and safety,*
- *Mass increases due to consumer driven features and content (autonomous driving, connectivity, smart mobility),*
- *Supply base infrastructure for new materials and processes, and*
- *Maintaining vehicle affordability.*

The MTT also recognized the opportunity offered by these challenges that will involve development of new vehicle architectures. These clean sheet approaches enable optimal use of new materials and full benefit from weight de-compounding.

The DOE-enacted budgets showed a significant decrease in funding for the materials area from \$29.0 million/\$28.5 million in FY 2014/2015 to \$21.6 million in FY 2016. It is unclear why this funding decrease is justified given the significant challenges identified by the MTT.

Response to Recommendations from the NRC Phase 4 Review

Two recommendations on materials were made in the NRC Phase 4 report (NRC, 2013, p. 110). The Partnership responses are shown below as "Partnership Responses" and the committee reaction as "Committee Assessment."

> ***NRC Phase 4 Recommendation 3-17.*** The Partnership should expand its current work on low-cost carbon-fiber precursors, manufacturing, and recycling. This work could also potentially help to reduce the cost of high-pressure hydrogen storage tanks. (Phase 3 Recommendation 3-24 urged the development of methods to recycle carbon-fiber composites.)
>
> ***Partnership Response.*** Clarification: the Partnership itself does not have a budget. As stated in the U.S. DRIVE Partnership Plan, each partner makes its own decisions regarding the funding and management of its projects.
>
> - The Partnership agrees that investments in carbon fiber and composites are important.
> - Within DOE's Office of Energy Efficiency and Renewable Energy (EERE), the Vehicle Technologies Office supports a number of ongoing projects in this area through its lightweight materials activity, and it coordinates closely with and leverages significant investments made by EERE's Advanced Manufacturing Office, as well as DOE Vehicle Technologies Recovery Act investments in carbon fiber. Projects are highlighted in the following slide.
> - Technical targets for the performance of carbon fiber for hydrogen storage cylinders are significantly more demanding than those for vehicle structural composites. However, within EERE, the Vehicle Technologies, Advanced Manufacturing, and Fuel Cell Technologies Offices closely coordinate their activities in carbon fiber and composites so each can leverage DOE investments for maximum benefit across the Department.

Committee Assessment of Response to 3-17. The committee strongly supports a balanced investment in the key lightweight materials (AHSS, Al, Mg and carbon-fiber composites). The MTT presentation made reference to 12 projects on composites which were cataloged in 2013 that address many of the challenges in using this material (see Table 3-21).

Another table that the committee received from DOE with U.S. DRIVE-related projects shows the projects listed in Table 3-22. This list reflects only 2 of the 12 projects listed by the MTT on composites which is clearly insufficient. DOE later clarified that this list reflects the 2015 AMR review projects. The 2016 AMR listed 4 of 19 projects on carbon-fiber composites which reflect an increasing emphasis on this material.

TABLE 3-21 Selections of Recent and Ongoing EERE Projects Related to Carbon Fiber and Composites

EERE Office	Project Lead (Team)	Project
Vehicle Technologies	ORNL	Plasma oxidation (co-funded MAP with the Advanced Manufacturing Office)
Vehicle Technologies	ORNL	Operating Funds: Carbon Fiber Technology Center
Vehicle Technologies	Zoltek with Weyerhauser	Low-cost carbon fiber
Vehicle Technologies	Material Innovation Technology Inc	Low-cost carbon fiber composites manufacturing Phase II SBIR
Vehicle Technologies	PNNL and ORNL (Moldflow)	Implementing predictive engineering models to commercial system
Vehicle Technologies	PNNL (Toyota)	Phase III: Predictive engineering complex 3-D shape validation
Vehicle Technologies	ORNL (Ford)	Phase III: Predictive engineering complex 3-D shape validation
Vehicle Technologies	USAMP	Crash model validation of carbon fiber composites
Advanced Manufacturing	ORNL	Operating Funds: Carbon Fiber Technology Center (co-funded with the Vehicle Technologies Office)
Advanced Manufacturing	Dow (Ford/ORNL)	Low-cost carbon fiber
Advanced Manufacturing	ORNL (Dow)	Low-cost carbon fiber
Vehicle Technologies (Recovery Act)	ORNL	Carbon Fiber Technology Center

NOTE: Acronyms defined in Appendix D.
SOURCE: Zaluzec and Joost (2016).

TABLE 3-22 Annual Merit Review (AMR) Project List: Vehicle Technologies Office, Lightweight Materials

Project ID	Presentation Title	Organization
lm094	Microstructure and the Corrosion/Protection of Cast Magnesium Alloys	Arizona State University
lm084 [a]	Validation of Material Models for Automotive Carbon Fiber Composite Structures	GM
lm087	Active, Tailorable Adhesives for Dissimilar Material Bonding, Repair, and Assembly	Michigan State University
lm095	A System Multiscale Modeling and Experimental Approach to Protect Grain Boundaries in Magnesium Alloys from Corrosion	Mississippi State University
lm086	Collision Welding of Dissimilar Materials by Vaporizing Foil Actuator	Ohio State University
lm093	High-Throughput Study of Diffusion and Phase Transformation Kinetics of Magnesium-Based Systems for Automotive Cast Magnesium Alloys	Ohio State University
lm076	Understanding Protective Film Formation by Magnesium Alloys in Automotive Applications	Oak Ridge National Laboratory
lm096	Corrosivity and Passivity of Metastable Magnesium Alloys	Oak Ridge National Laboratory
lm057	Mechanistic-Based Ductility Prediction for Complex Magnesium Castings	Pacific Northwest National Laboratory
lm074	SPR Process Simulation, Analyses, and Development for Magnesium Joints	Pacific Northwest National Laboratory
lm079	Enhanced Room-Temperature Formability in High-Strength Aluminum Alloys through Pulse-Pressure Forming	Pacific Northwest National Laboratory
lm092	In Situ Investigation of Microstructural Evolution during Solidification and Heat Treatment in a Die-Cast Magnesium Alloy	Pacific Northwest National Laboratory
lm099[a]	High-Strength, Dissimilar Alloy Aluminum Tailor-Welded Blanks	Pacific Northwest National Laboratory
lm091	Phase Transformation Kinetics and Alloy Microsegregation in High-Pressure Die Cast Magnesium Alloys	University of Michigan
lm080 [a]	Integrated Computational Materials Engineering Approach to Development of Lightweight 3GAHSS Vehicle Assembly	United States Automotive Materials Partnership

TABLE 3-22 Continued

Project ID	Presentation Title	Organization
lm081	GATE Center of Excellence at UAB for Lightweight Materials and Manufacturing for Automotive, Truck and Mass Transit	University of Alabama, Birmingham
lm072[a]	Multi-Material Lightweight Vehicles	Vehma
lm089	High-Strength Electroformed Nanostructured Aluminum for Lightweight Automotive Applications	Xtalic Corporation
lm100	Upset Protrusion Joining Techniques for Joining Dissimilar Metals	Fiat Chrysler Automobiles US LLC
lm035	Scale-Up of Magnesium Production by Fully Stabilized Zirconia Electrolysis	INFINIUM, Inc.
lm098	Brazing Dissimilar Metals with a Novel Composite Foil	Johns Hopkins University
lm006[a]	Advanced Oxidation and Stabilization of PAN-Based Carbon Precursor Fibers	Oak Ridge National Laboratory
lm097	Laser-Assisted Joining Process of Aluminum and Carbon Fiber Components	Oak Ridge National Laboratory
lm078	Aluminum Formability Extension through Superior Blank Processing	Pacific Northwest National Laboratory
lm077	Magnesium-Intensive Front End Sub-Structure Development	United States Automotive Materials Partnership

[a] Presented at committee meeting on June 22, 2016.

NOTE: The color coding in the table was done by the committee: yellow, magnesium (10 of 24); blue, aluminum (4 of 24); green, carbon fiber (2 of 24); gray, steel (1 of 24); orange, joining (5 of 24); white, other (2 of 24).

SOURCE: Project numbers are from the Department of Energy's 2016 Annual Merit Review, see Cooper (2016a).

The committee continues to endorse a balanced portfolio of advanced high-strength steel (AHSS), Mg, Al, and carbon-fiber composites that also include an increased emphasis on joining these materials. Note that the 2016 AMR had 7 of 19 projects on joining compared to 5 of 24 projects in the 2015 AMR.

> ***NRC Phase 4 Recommendation 3-18.*** The materials technical team should expand its outreach to the other technical teams to determine the highest priority collective Partnership needs, and the team should then reassess its research portfolio accordingly. Any necessary reallocation of resources could be enabled by delegating some of the highly competitive metals development work to the private sector.
>
> ***Partnership Response.*** USDRIVE agrees that outreach to other technical teams is important to leverage common materials issues.

- At our joint all tech team meeting in Oct 2011 USDRIVE explored the possibility of a new "Advanced Materials Tech Team" including gathering information from other Tech Teams relative to their material needs. However, USDRIVE determined that each team has unique needs, and generally has access to the needed experts who are actively engaged, but are limited by funding availability.
- Several DOE VTO lightweighting projects leverage other DOE programs. For example, in carbon fiber composites the hydrogen storage program builds upon the work funded both by VTO and AMO.
- The MTT emphasizes lightweighting structural materials facing significant technical hurdles which are unique to each of those materials. Since light weighting serves as an enabler for improving vehicle efficiency, and reducing the size and mass of propulsion systems regardless of which advanced propulsion system is employed, we feel it is appropriate to continue funding the precompetitive research in an array of structural materials, rather than diverting funding to other technologies.

Committee Assessment of Response to 3-18. The MTT presented a summary of their interaction and leveraging with the other technical teams. The committee endorses the approach taken by the MTT. Some of the issues related to materials within the other technical teams are as follows (Zaluzec and Joost, 2016):

- Safety
  - —Low-cost, high-volume, industrial-grade carbon fiber.
  - —Joint collaboration on carbon fiber reinforced composites including development materials cards [data] and crash computer aided engineering (CAE) models.
- Power trains
  - —Material concerns for power train include increased temperatures and pressures for boosted high-performance engines.
- Energy storage
  - —Materials and process information on low-cost carbon fiber for Type IV storage tanks.
- Manufacturing
  - —Multimaterial vehicles (cost-effective weight savings).
  - —Design and architectures requires manufacturing support (joining and assembly).
  - —MTT continues expansion into chassis and power train.

This committee also endorses the position taken by the MTT that the existing and future projects in the materials roadmap are precompetitive in nature and a priority to achieving the targeted weight reduction. Therefore, it is not appropriate to reallocate the resources applied to this key technology area.

Appropriate Federal Role

It is entirely appropriate for the federal government to be involved in precompetitive research that addresses the underlying issues toward achieving vehicle weight reduction. The target of 50 percent weight reduction requires technology breakthroughs in material composition and processing that require deep experimental and modeling research. The federal funding that supports the unique capabilities at universities and national laboratories is key in this regard.

Further, U.S. DRIVE provides a mechanism for integrating the activities across the breadth and complexity of the automotive materials/component supply chain. Coordinating and integrating the activities across the materials producers, component manufacturers, and OEMs is critical to achieving the required fuel economy improvements enabled by weight reduction.

Findings and Recommendations

Finding 3-40. The U.S. DRIVE materials technical team (MTT), like the FreedomCAR and Fuel Partnership before it, has adopted a stretch goal of 50 percent reduction in vehicle weight (versus 2002 comparable vehicles) *with equal affordability* (emphasis added). Previous NRC committees found this goal to be unrealistic. The present committee agrees with that assessment. Further, during this review, the MTT reported a revision to the target by setting the comparator as a 2015 vehicle. The committee feels that this makes the target even more unrealistic.

Finding 3-41. The committee was pleased to see the MTT reported adoption of a midterm target for 2020 of an 18 percent weight glider reduction to be achieved at <$11/kg (<$5/pound) saved while maintaining equal vehicle-level performance (crash, nvh, durability, reliability, and recyclability).[24] However, the 2016 DOE Annual Merit Review referred to a 30 percent reduction by 2022 relative to a 2012 baseline (Wu, 2016). This is not realistic given the 2020 target. When designing a new vehicle, the options available for weight reduction are compared with other fuel economy improvement solutions including increased power train efficiency, vehicle electrification, decreased tire rolling resistance, and improved aerodynamics. It appears to be more appropriate to set a long-term target for weight reduction for the Partnership, in a similar manner.

Recommendation 3-29. U.S. DRIVE should set the long-term target for the cost of weight reduction to be consistent with the long-term cost targets for the other technical teams. The committee also recommends continuing the practice of setting midterm targets. In doing so, it is important for all Department of Energy and U.S. DRIVE sources to reference a consistent set of targets.

Recommendation 3-30. U.S. DRIVE should set a new basis for the cost targets that captures the full life cycle including end-of-life and repair costs, and the

---

[24] The glider is the vehicle structure excluding the power train. Historically weight reductions are easier to achieve in that part of the vehicle.

carbon footprint, when looking at incremental cost of weight savings by material substitution. This should be done in collaboration with the cradle-to-grave working group.

Finding 3-42. The Multi-Material Lightweight Vehicle (MMLV) project demonstrated the increased weight savings that are possible when the best material is used in each component. This requires that all four of the material options (AHSS, Al, Mg and carbon-fiber composite) are developed in parallel together with new multimaterial joining processes. It is important to achieve a portfolio that can address the technology gaps across all the materials. This is challenging given a materials portfolio that was reduced from $29.0 million in FY 2014 to $21.6 million in FY 2016.

Recommendation 3-31. The lightweighting materials project portfolio should correctly reflect what is needed to meet the challenges of achieving the target of affordable vehicle weight reduction.

CRADLE-TO-GRAVE ANALYSIS AND IMPLICATIONS

The NRC Phase 4 report recognized based on Phases 1, 2, and 3 reports that "it is critical to understand the environmental implications of the full life cycle of alternative fuel pathways, including hydrogen, electricity, biofuels, or other energy source/vehicle combinations being developed that can potentially reduce the consumption of petroleum and reduce greenhouse gas emissions relative to conventional light-duty vehicles" (2013). In fact, the recommendation for life cycle analysis of vehicle-fuel pathways, often termed C2G analysis, was made in the NRC Phase 3 report: "The Partnership should consider incorporating the broader scope of 'cradle to grave' analysis rather than a 'source (well)-to-wheels' approach in program planning from production to recycling in order to better consider total energy consumption, total emissions, and the total environmental impact of various energy/vehicle pathways and technologies" (NRC, 2010).

Organization of the Task

This C2G project and analysis effort was organized as a U.S. DRIVE working group as distinct from the technical teams (see Figure 2-1). Working groups are ad hoc teams, assigned to accomplish a specific task with the implication that the working group can then disband once the task is complete.

The C2G working group recommended the analytic approach taken to implement the NRC Phase 3 recommendation and the analytical work accomplished by the ANL, which is published in Elgowainy et al. (2016).

About the Argonne National Laboratory Report

The ANL report offers a C2G analysis of the cost and GHG emissions from the principal vehicle-fuel pathways, as well as the levelized cost of driving and cost of avoided GHG emissions. Only GHG emissions were modeled. Other externalities such as air quality, and considerations of vehicle functionality (range, refueling time and infrastructure availability, packaging), and fuel production scalability, are not captured in this study.

This was accomplished chiefly through use of the ANL GREET model, whose full name implies the functional description: *G*reenhouse Gases, *R*egulated *E*missions, and *E*nergy Use in *T*ransportation. Figure 3-20 shows items in the vehicle cycle GREET considered: raw material extraction, material processing, component manufacture and vehicle assembly, vehicle operation, and vehicle end of life. The fuel cycle is also shown: raw material extraction, transportation, refining, and delivery. The unique contribution of this project derives from the way the model was used. The analysis combines two streams of input, the GHG emissions from the fuel cycle and the GHG emissions from the production of the vehicle itself, to yield an overall figure of merit for each vehicle-fuel pathway. The results of any selected pathway can be compared with any other due to a framework of common assumptions and analytic treatment.

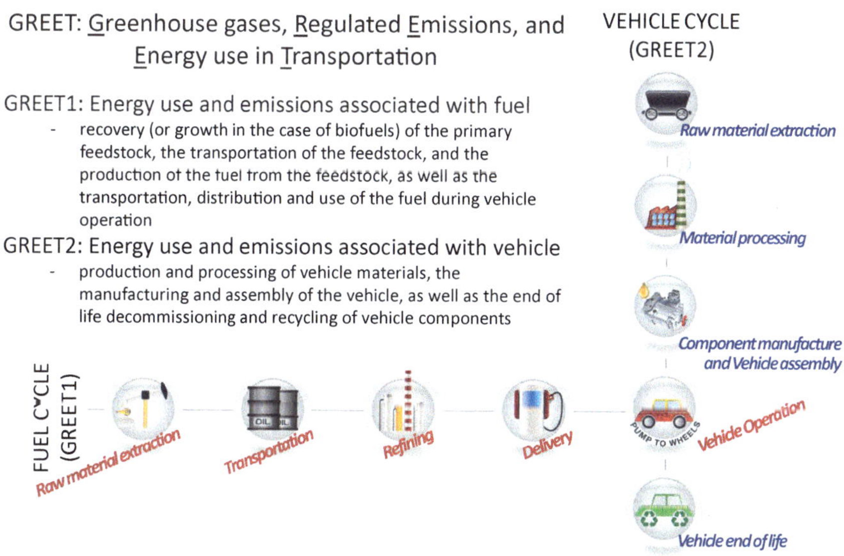

FIGURE 3-20 Schematic description of the full fuel cycle analysis of the Argonne National Laboratory GREET (Greenhouse Gases, Regulated Emissions, and Energy Use in Transportation) model. SOURCE: Joseck and Ward (2016).

The term "pathway" as used in the ANL report deserves special mention. This refers to a defined and unique set of fueling options that is uniquely matched with and supports each selected vehicle type.[25] The analysis does not include considerations of macroeconomic conditions, assumed policy choices, and market desirability, nor does it allow a blend of pathways. Thus, each pathway illustrates the distinct consequences and costs that would occur if that pathway became the actual outcome. Finally, the analysis includes only those pathways for which the technologies are either in use now or for which the technology readiness level can enable future commercial development.

While thorough and robust, the C2G model remains costly and complex to run. As a consequence, it would be useful for the U.S. DRIVE C2G working group to construct a spreadsheet model that would be simpler to use and not always require running the larger GREET model at ANL to address "what if" policy questions.

Principal Conclusions from the Analysis

The modeling work provided a rich set of alternatives and insights. This is illustrated with some examples from the report. All these results were measured against a baseline of current vehicle performance, 26.2 miles per gallon typical of vehicles like the Chevrolet Malibu, the Dodge Dart, or the Ford Fusion. Internal combustion vehicles running on gasoline developed from pyrolysis of forest residues were modeled to have C2G GHG emissions of about 140 g carbon dioxide equivalents (CO_2-eq)/mi. HFCVs running on hydrogen produced from biomass gasification would have emissions of about 115 g CO_2-eq/mi. Battery electric vehicles running on wind-generated electricity and HFCVs running on hydrogen from wind-generated electricity would have C2G GHG emissions of about 50 g CO_2-eq/mi or less. In contrast, a contemporary internal combustion engine vehicle running on gasoline produces about 450 g CO_2-eq/mi. The cradle-to-grave analysis also illustrates how the source of electric energy influences the total emissions that can be attributed to any electric vehicle. For example, Figure 3-21 shows that a BEV with a range of 90 miles would result in only 50 grams of CO_2 per mile if the energy were generated by wind. In contrast, the same vehicle would cause about 180 grams CO_2 per mile if the generation source were coal.

Conclusions like this can be compared systematically through summary graphics like those shown in Figures 3-21 and 3-22. In Figure 3-21, the vertical axis measures the GHG emissions in grams of CO_2-eq per mile of vehicle travel.

[25] More formally, the C2G working group defines the fuel "pathway" as follows: "A fuel or energy production pathway is defined as a distinct, technically feasible, route or a sequence of processes starting with one or more feedstocks and ending with an intermediate or a final product. A pathway is not necessarily constrained by feedstock, economic, policy, and market considerations" (Joseck and Ward, 2016).

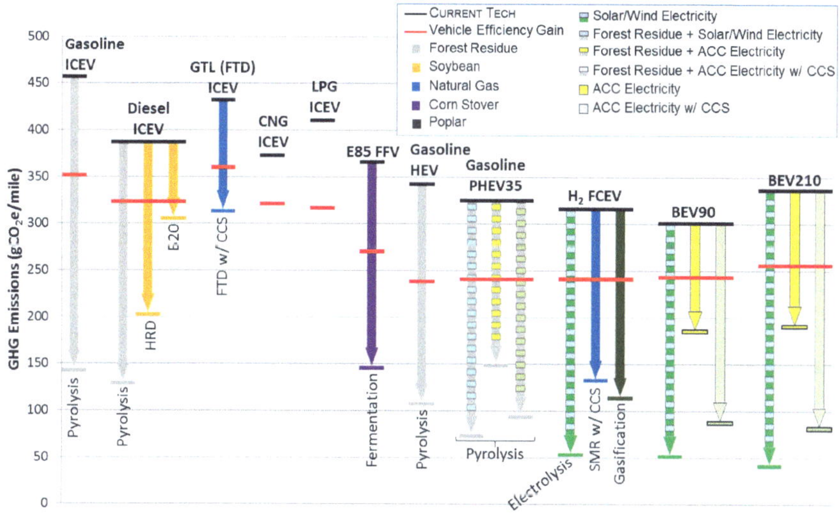

FIGURE 3-21 Results of the full fuel-cycle analysis of greenhouse gas emissions for different energy-vehicle combinations. SOURCE: Elgowainy et al. (2016).

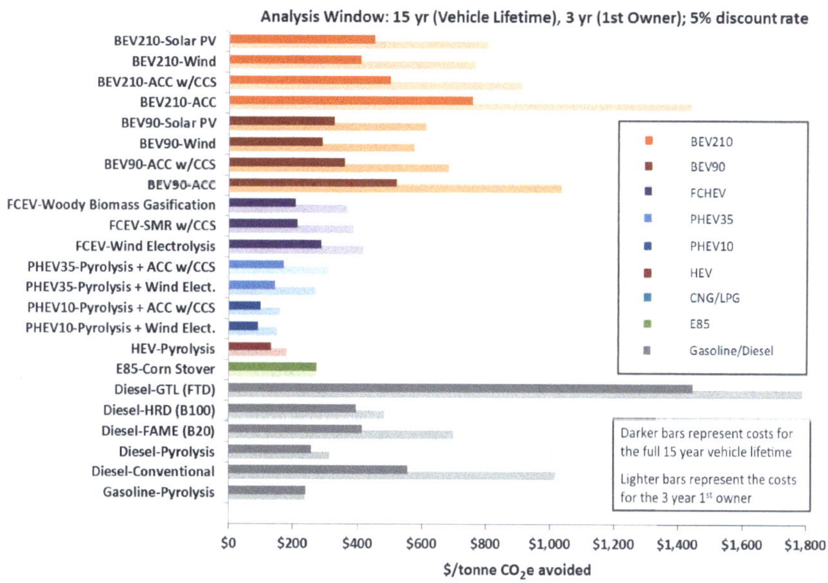

FIGURE 3-22 Results of the full fuel-cycle analysis of the cost of avoided greenhouse gas emissions for different energy-vehicle combinations. SOURCE: Elgowainy et al. (2016).

The various vehicle-fuel pathways appear along the horizontal axis.[26] The vertical, downward-pointing arrows summarize the estimates of the ANL analysts and the U.S. DRIVE C2G working group concerning the possibility of improvement for each pathway.

However, none of these remedies is without cost, and the report allows cost comparisons among the vehicle-fuel pathways in terms of dollars per metric ton of CO_2-eq avoided (in this case, based on the projections of the progress achieved by future technologies). Figure 3-22 illustrates these results with a measure of the cost of GHGs avoided.

The ANL study reached the following conclusions, which were stated as shown here in the presentation to the committee at its June 22-23, 2016, meeting (Joseck and Ward, 2016):

- Emissions
  - —Large GHG reductions for light-duty vehicles are challenging and require consideration of the entire life cycle, including vehicle manufacture, fuel production, and vehicle operation.
- Cost
  - —High-volume production is critical to the viability of advanced technologies.
  - —Incremental costs of advanced technologies in FUTURE TECHNOLOGY, HIGH VOLUME cases are significantly reduced, reflecting estimated R&D outcomes.
  - —Low-carbon fuels can have significantly higher costs than conventional fuels.
  - —Vehicle cost is the major (60 to 90 percent) and fuel cost the minor (10 to 40 percent) component of the levelized cost of driving when projected at volume. Treatment of residual vehicle cost is an important consideration. Many alternative vehicles and/or fuels cost significantly more than conventional gasoline vehicles for the CURRENT TECHNOLOGY case, even when costs are projected for high-volume production.
- Cost of Carbon Abatement
  - —For the CURRENT TECHNOLOGY, HIGH VOLUME case, carbon abatement costs are generally on the order of \$100s per tonne CO_2 to \$1,000s per tonne CO_2 for alternative vehicle-fuel pathways compared to a conventional gasoline vehicle baseline.
  - —FUTURE TECHNOLOGY, HIGH VOLUME carbon abatement costs are generally expected to be in the range \$100 to \$1,000/tonne CO_2.

[26] A table of "Acronyms and Initialisms" appears on page xiii of Elgowainy et al. (2016) and is essential for decoding the vehicle-fuel pathways in the graphic.

- Technology Feasibility
  - Significant technical barriers still exist for the introduction of some alternative fuels. Further, market transition barriers—such as low-volume cost, fuel, or make/model availability, and vehicle/fuel/infrastructure compatibility—may play a role as well.

Response to Recommendations from the NRC Phase 4 Review

The NRC Phase 4 report made a couple of recommendations related to life-cycle analysis and the Partnership provided responses to them as follows:

> ***NRC Phase 4 Recommendation 2-10.*** The U.S. DRIVE Partnership should integrate a life-cycle assessment approach into its research portfolio for energy storage batteries, fuel cell stacks, power electronics, hydrogen fuel tanks, and other advanced vehicle components in order to gain an understanding of the potential environmental impacts of materials processing, supply chains, manufacture, and vehicle use and end of life. U.S. DRIVE should anticipate the potential risk and environmental externalities of battery production and end of life and should research methods to minimize these impacts.
>
> ***Partnership Response.*** The Partnership agrees with this recommendation and—based on Executive Steering Group direction in June 2011 and subsequent discussion at the October 2011 U.S. DRIVE All Technical Team Meeting—established a C2G Working Group in January 2012, drawing from expertise in the Vehicle Systems and Analysis and Fuel Pathway Integration Technical Teams, as well as other analytical expertise from U.S. DRIVE partner organizations. As part of this effort, Argonne National Laboratory (ANL) updated the GREET (Greenhouse Gases, Regulated Emissions, and Energy Use in Transportation) model's analysis capabilities to include lifecycle assessment of fuel cell vehicle and hydrogen manufacturing. With input from the working group, ANL also examined material and energy flows for lithium-ion batteries, and data relevant to active cathode and anode materials, battery electronics, battery assembly, and battery recycling were incorporated in the GREET model for environmental lifecycle assessment of plug-in hybrid and battery electric vehicles. The working group has performed lifecycle assessments of multiple vehicle pathways and compared the lifecycle greenhouse gas emissions on a "per mile" basis. ANL reported results of the lifecycle assessment in the 2012 DOE Annual Merit Review and Peer Evaluation Report, and the working group presented its initial results to the ESG in October 2012 as well as an update in June 2013.

Committee Assessment of Response to 2-10. The committee recognizes and appreciates the engagement of the Executive Steering Group and the senior leaders of the Partnership in creating this useful management and policy tool. The committee's recommendations for the continued growth and improvement of this policy tool appear at the end of this section.

> ***NRC Phase 4 Recommendation 2-11.*** The Executive Steering Group as well as the systems analysis teams of the U.S. DRIVE Partnership should identify pathways for fuel cell vehicles and electric vehicles to achieve large life-cycle greenhouse gas (GHG) reductions and structure risk-weighted R&D portfolios to increase the likelihood of achieving these goals at competitive costs. U.S. DRIVE should also update and publicly publish

comparisons of per mile life-cycle GHG emissions across vehicle technologies regularly so that stakeholders can understand all assumptions made, be aware of systems impacts, and identify potential improvements.

Partnership Response. The Partnership agrees with this recommendation. As noted in the response to Recommendation 2-10 above, the Partnership established a Cradle-to-Grave (C2G) Working Group to identify lifecycle assessment analytical gaps, perform lifecycle assessments of multiple vehicle pathways, and compare the lifecycle GHG emissions on a "per mile" basis across pathways. In support of this activity, the U.S. Department of Energy funded Argonne National Laboratory (ANL) to update the GREET (Greenhouse Gases, Regulated Emissions, and Energy Use in Transportation) model's analysis capabilities to include lifecycle assessment of vehicle and manufacturing cycles, and the C2G Working Group is using ANL's GREET model to perform these assessments and analyses. The C2G Working Group prepared a preliminary lifecycle assessment and internal presentation for the ESG of 10 technology and vehicle pathways. The assessment identified GHG emissions on a "per mile" basis for each pathway, including the contributions from the fuel, vehicle, and manufacturing cycles, and examined the GHG emissions of the battery manufacturing portion of the vehicle lifecycle. The C2G Working Group plans to continue detailed assessments and uncertainty analysis of the GHG emissions and identify potential improvements and, per ESG direction, plans to make its work publicly available via the DOE website.

Committee Assessment of Response to 2-11. The committee appreciates the prompt and highly capable management and policy tool that has been created. The committee further notes that publishing the results in the open literature will enable independent analysts and researchers to contribute to the larger body of knowledge concerning vehicle and fuel pathways. The committee's recommendations for the ongoing work to continue development of this management and policy tool appear below.

Findings and Recommendations

Finding 3-43. The cradle-to-grave life-cycle analysis model provides a major step forward in the ability of the U.S. DRIVE to advise the industry and the Department of Energy on program and policy choices. This tool provides the capabilities and insights that will give the Partnership a useful management tool and with further development, a strategic and policy capability.

Recommendation 3-32. The cradle-to-grave model should be continually updated and, where possible, tailored to improve its ability to support senior policy makers. Resources appropriate to this task should be provided. This updating will be an ongoing project, and the Partnership should consider upgrading the ad hoc working group to a technical team.

Finding 3-44. The next step in the growth of the cradle-to-grave model will be to improve its relevance to the senior leadership of the DOE and the road mobility industry in managing their policy choices and program investments.

Recommendation 3-33. As the cradle-to-grave model is improved, interaction with senior leaders at the Executive Steering Group and the Joint Operations Group should be emphasized as a source of guidance that will ensure policy relevance of the model. Most immediately, the model could be extended to include the impact of vehicle weight reduction.

Finding 3-45. Creation of the cradle-to-grave model has already shown value across the Partnership. The process of building the model required a harmonization of assumptions across the many technology teams within the U.S. DRIVE Partnership, thus facilitating comparability among work products.

Recommendation 3-34. The cradle-to-grave model should be improved:

- Assumptions regarding the progress of vehicle technologies should be reviewed and updated, especially those for the battery electric vehicle.
- Development of a simplified and policy-relevant spreadsheet model should be done in parallel with the updating/improvement on the basic GREET-based model to ensure that the two remain compatible. The intent of the spreadsheet model should be to support policy development and strategic program decision making by senior leaders in industry and government.

REFERENCES

Abengoa Bioenergy. 2014. "2G Hugoton Project: General Information." http://www.abengoabioenergy.com/web/en/2g_hugoton_project/.

Ainscough, C., D. Peterson, and E. Miller. 2014. "Hydrogen Production Cost from PEM Electrolysis." *DOE Hydrogen and Fuel Cells Program Record 14004*, July 1. https://www.hydrogen.energy.gov/pdfs/14004_h2_production_cost_pem_electrolysis.pdf.

Allendorf, M., and R. Borup. 2016. FCTO Consortia Overview Webinar (HyMARC and FC-PAD). U.S. Department of Energy Fuel Cells Technologies Office, January 7. http://energy.gov/eere/fuelcells/events/webinar-fcto-consortia-overview-webinar-fc-pad-and-hymarc.

ARPA-E (Advanced Research Projects Agency-Energy). 2012. "MOVE, Methane Opportunities for Vehicular Energy." https://arpa-e.energy.gov/?q=arpa-e-programs/move.

Brooke, L., R. Gehm, and B. Visnic. 2016. Lightweighting—What's next?, *Automotive Engineering* August:16-25. http://www.sae.org.cn/uploads/pubs/magazines/2017/AEI/sae_16AUTP08.pdf.

Brown, R., and S. Capareda. 2015. Producing biofuels via the thermochemical platform. *CEP* March: 41-44.

Burress, T. 2016. "Advanced Electric Motor Research." Oak Ridge National Laboratory, Project EDT062, Presentation to the 2016 U.S. Department of Energy Annual Merit Review, Washington, D.C., June 7. http://energy.gov/sites/prod/files/2016/06/f32/edt062_burress_2016_o_web.pdf.

Casady, J. 2016. "88 Kilowatt Automotive Inverter with New 900 Volt Silicon Carbide MOSFET Technology." Cree, Inc., Project EDT073, Presentation to the 2016 U.S. Department of Energy Annual Merit Review, Washington, D.C., June 8. http://energy.gov/sites/prod/files/2016/06/f32/edt073_casady_2016_o_web.pdf.

Chadwell, C., T. Alger, J. Zuehl, and R. Gukelberger. 2014. A demonstration of dedicated EGR on a 2.0 L GDI engine. *SAE Int. J. Engines* 7(1):434-447.

Chapman, B., and K. Randolph. 2016. "Hydrogen Production Tech Team. Presentation to the NRC Committee." Presentation to the Committee on the Review of the Research Program of the U.S. DRIVE Partnership, Phase 5, Washington, D.C., April 19.

Chinthavali, M. 2016. "Electric Drive Inverter R&D." Oak Ridge National Laboratory, Project EDT053, Presentation to the 2016 U.S. Department of Energy Annual Merit Review, Washington, D.C., June 8. http://energy.gov/sites/prod/files/2016/06/f32/edt053_chinthavali_2016_o_web.pdf.

Cooper, C. 2016a. FCTO VTO AMR Project List for U.S. DRIVE. E-mail communication to the committee (pers. comm.), March 4.

Cooper, C. 2016b. "U.S. DRIVE Overview Presentation." Presentation to the Committee on the Review of the Research Program of the U.S. DRIVE Partnership, Phase 5, Washington, D.C., February 3.

CryoGas International. 2016. Liquid hydrogen storage system. *Hydrogen Issue* 54(7):6.

Daw, C.S., R.L. Graves, R.M. Wagner, and J.A. Caton. 2010. *Report on the Transportation Combustion Engine Efficiency Colloquium Held at USCAR, March 3-4, 2010.* http://info.ornl.gov/sites/publications/files/Pub26829.pdf.

Dawsey, R., and S. Rogers. 2016. "Electrical and Electronics Technical Team." Presentation to the Committee on the Review of the Research Program of the U.S. DRIVE Partnership, Phase 5, Detroit, June 22.

DeVoto, D. 2016. "Performance and Reliability of Bonded Interfaces for High-Temperature Packaging." National Renewable Energy Laboratory, Project EDT063, Presentation to the 2016 U.S. Department of Energy Annual Merit Review, Washington, D.C., June 8. http://energy.gov/sites/prod/files/2016/06/f32/edt063_devoto_2016_o_web.pdf.

DOE (U.S. Department of Energy). 2010. *2010 Annual Progress Report: Lightweighting Material.* http://www1.eere.energy.gov/vehiclesandfuels/pdfs/program/2010_lightweighting_materials.pdf.

DOE. 2014a. *Energy Storage Annual Report.* Washington, D.C.: U.S. Department of Energy Vehicle Technologies Office. http://energy.gov/eere/vehicles/downloads/vehicle-technologies-office-2014-energy-storage-rd-annual-report.

DOE. 2014b. *EV Everywhere Grand Challenge: Road to Success.* Washington, D.C.: U.S. Department of Energy. http://energy.gov/sites/prod/files/2016/05/f31/eveverywhere_road_to_success.pdf.

DOE. 2015a. *Energy Storage Annual Report.* Washington, D.C.: U.S. Department of Energy Vehicle Technologies Office. http://energy.gov/eere/vehicles/downloads/vehicle-technologies-office-2015-energy-storage-rd-annual-report.

DOE. 2015b. *Quadrennial Technology Review: An Assessment of Energy Technologies and Research Opportunities.* http://energy.gov/sites/prod/files/2015/09/f26/Quadrennial-Technology-Review-2015_0.pdf.

DOE. 2016a. "Alternative Fuels Data Center: Availability of Hybrid and Plug-In Electric Vehicles." http://www.afdc.energy.gov/vehicles/electric_availability.html. Accessed November 7, 2016.

DOE. 2016b. "Alternative Fuels Data Center: Maps and Data." http://www.afdc.energy.gov/data/. Accessed September 13, 2016.

DOE. 2016c. "Co-Optimization of Fuels and Engines." http://energy.gov/eere/bioenergy/co-optimization-fuels-engines. Accessed October 26, 2016.

DOE. 2016d. "Energy Department Announces $7 Million to Accelerate Fuel and Engine Co-Optimization Technologies." https://energy.gov/eere/articles/energy-department-announces-7-million-accelerate-fuel-and-engine-co-optimization.

DOE. 2016e. *Pathways to Commercial Success: Technologies and Products Supported by the Fuel Cell Technologies Office.* http://energy.gov/sites/prod/files/2016/02/f29/fcto_2015_pathways_commercial_success.pdf.

DOE. 2016f. "Vehicle Technologies Office: Annual Merit Review Presentations." http://energy.gov/eere/vehicles/vehicle-technologies-office-annual-merit-review-presentations. Accessed September 20, 2016.

Edwards, D. 2016. "Accelerating Electric Drives: The Next Generation of Hydrogen Fuel Cells." Presentation to the Committee on the Review of the Research Program of the U.S. DRIVE Partnership, Phase 5, Detroit, June 22.

EIA (Energy Information Administration). 2016a. *Annual Energy Outlook 2016 Early Release: Annotated Summary of Two Cases.* http://www.eia.gov/forecasts/aeo/er/pdf/0383er(2016).pdf.

EIA. 2016b. "What Is U.S. Electricity Generation by Energy Source?" https://www.eia.gov/tools/faqs/faq.cfm?id=427&t=3. Accessed September 10, 2016.

El-Refaie, A. 2016. "Alternative High-Performance Motors with Non-Rare Earth Materials." GE Global Research, Project EDT045, Presentation to the 2016 U.S. Department of Energy Annual Merit Review, Washington, D.C., June 7. http://energy.gov/sites/prod/files/2016/06/f32/edt045_elrefaie_2016_o_web.pdf.

Elgowainy, A., J. Han, J. Ward, F. Joseck, D. Gohlke, A. Lindauer, T. Ramsden, M. Biddy, M. Alexander, S. Barnhart, I. Sutherland, L. Verduzco, and T.J. Wallington. 2016. *Cradle-to-Grave Lifecycle Analysis of U.S. Light-Duty Vehicle-Fuel Pathways: A Greenhouse Gas Emissions and Economic Assessment of Current (2015) and Future (2025-2030) Technologies.* Argonne National Laboratory (ANL/ESD-16/7), June. https://greet.es.anl.gov/publication-c2g-2016-report.

EPA (U.S. Environmental Protection Agency). 2014. *Light-Duty Automotive Technology, Carbon Dioxide Emissions, and Fuel Economy Trends: 1975 Through 2014—Executive Summary.* EPA Publication 420-S-14-001, October. http://www.epa.gov/fueleconomy/fetrends/1975-2014/420s14001.pdf.

Farenback-Brateman, J., K. Stork, and W. Studzinski. 2016. "Fuels Working Group (FWG) Presentation to the National Research Council Committee on the Review of the U.S. Drive Partnership." Presentation to the Committee on the Review of the Research Program of the U.S. DRIVE Partnership, Phase 5, Detroit, June 22.

Farrell, J. 2016. "Co-optimization of Fuels and Engines." SAE High Efficiency Internal Combustion Engine Symposium, Detroit, April 11. http://energy.gov/sites/prod/files/2016/07/f33/farrell_co-optimization_sae_heice_%20symposium_2016.pdf.

Fenske, G., L. Ajayi, N. Demas, R. Erck, C. Lorenzo-Martin, A. Erdemir, and O. Eryilmaz. 2015. "Engine Friction Reduction Technologies." Argonne National Laboratory, Project FT012, Presentation at the 2015 U.S. Department of Energy Annual Merit Review, Washington, D.C., June 11. http://energy.gov/sites/prod/files/2015/06/f24/ft012_fenske_2015_o.pdf.

Forfia, D., M. Knight, and R. Melton. 2016. The view from the top of the mountain: Building a community of practice with the GridWise Transactive Energy Framework. *IEEE Power and Energy Magazine* 14(3):25-33. doi: 10.1109/MPE.2016.2524961.

General Motors. 2014. "GM Fuel Cell Fleet Tops 3 Million Miles: Chevrolet Equinox Fuel Cell Vehicles Driven in Real World Reach Milestone." Press Release, May 7. http://media.gm.com/media/us/en/gm/news.detail.html/content/Pages/news/us/en/2014/May/0507-fuel-cell.html.

General Motors. 2016. "GM Hydrogen Fuel Cells Mark 50 Years of Development: Electrovan Was First Fuel Cell Technology Transfer from JFK's Moonshot Challenge." Press Release, October 5. http://www.gm.com/mol/m-2016-oct-1005-hydrogen.html.

Green Car Congress. 2016. "SAE Technical Experts: Fuel Cell Technology Has Advanced Significantly, FC Vehicle Production Has Begun, Further Cost Reductions & Infrastructure Development Required." April 28. http://www.greencarcongress.com/2016/04/20160425-pfl720.html.

Greimel, H. 2016. Honda, GM set targets for 2020 fuel cell vehicle Clarity, a steppingstone, begins leasing in Japan. *Automotive News*, March 14. http://www.autonews.com/article/20160314/OEM05/303149972/honda-gm-set-targets-for-2020-fuel-cell-vehicle.

GridWise Architecture Council. 2015. *Transactive Energy Framework: Version 1.0.* http://www.gridwiseac.org/pdfs/te_framework_report_pnnl-22946.pdf.

Gupta, E., and H. Soto. 2016. "Hydrogen Delivery Technical Team (HDTT). Presentation to National Research Council." Presentation to the Committee on the Review of the Research Program of the U.S. DRIVE Partnership, Phase 5, Washington, D.C., April 19.

Holladay, J. 2016. "MagnetoCaloric Hydrogen Liquefaction." Pacific Northwest National Laboratory, Project PD131, Presentation to the 2016 U.S. Department of Energy Annual Merit Review, Washington, D.C., June 8.

Howell, D. 2016. "DOE/EERE Vehicle Technologies Office Overview." Presentation to the Committee on the Review of the Research Program of the U.S. DRIVE Partnership, Phase 5, Washington, D.C., February 3.

Howell, D., and R. Elder. 2012. "Electrochemical Energy Storage Technical Team (EEST)." Presentation to the Committee on the Review of the Research Program of the U.S. DRIVE Partnership, Phase 4, Washington, D.C., January 26.

Howell, D., and K. Snyder. 2016. "Electrochemical Energy Storage Technical Team (EEST)." Presentation to the Committee on the Review of the Research Program of the U.S. DRIVE Partnership, Phase 5, Washington, D.C., April 19.

IEA (International Energy Agency). 2016. *Global EV Outlook 2016: Beyond One Million Electric Cars.* https://www.iea.org/publications/freepublications/publication/Global_EV_Outlook_2016.pdf.

James, B.D., and A.B. Spisak. 2012. *Mass Production Cost Estimation of Direct H_2 PEM Fuel Cell Systems for Transportation Applications: 2012 Update, Revision 4.* https://www1.eere.energy.gov/hydrogenandfuelcells/pdfs/sa_fc_system_cost_analysis_2012.pdf.

James, B.D., J.M. Moton, D.A. DeSantis, and G. Saur. 2015. "Analysis of Advanced H2 Production Pathways." Strategic Analysis and National Renewable Energy Laboratory, Project PD102, Presentation to the 2015 U.S. Department of Energy Annual Merit Review, Washington, D.C., June 11.

Johnson, T. 2016. Vehicular emissions in review. *SAE International Journal of Engines* 9(2):1258-1275.

Joost, W.J. 2012. Reducing vehicle weight and improving U.S. energy efficiency using integrated computational materials engineering. *JOM* 64(9):1032-1038.

Joseck, F. 2016. "Systems Analysis Program Area." Plenary Presentation at the U.S. Department of Energy Annual Merit Review, Washington, D.C., June 7.

Joseck, F., and L. Verduzco. 2016. "Fuel Pathways Integration Tech Team Presentation to the NRC Committee." Presentation to the Committee on the Review of the Research Program of the U.S. DRIVE Partnership, Phase 5, Washington, D.C., April 19.

Joseck, F., and J. Ward. 2016. "Life-cycle Analysis of U.S. DRIVE Light-Duty Vehicle-Fuel Pathways." Presentation to the Committee on the Review of the Research Program of the U.S. DRIVE Partnership, Phase 5, Detroit, June 22.

Kristov, L., P. De Martini, and J. Taft. 2016. A tale of two visions: Designing a decentralized transactive electric system. *IEEE Power and Energy Magazine* 14(3):63-69. doi: 10.1109/MPE.2016.2524964.

Kurtz, J., S. Sprik, C. Ainscough, and G. Saur. 2015. "Fuel Cell Electric Vehicle Evaluation." National Renewable Energy Laboratory, Project TV001, Presentation at the 2015 U.S. Department of Energy Annual Merit Review, Washington, D.C., June 10.

Lane, J. 2014. Update from DuPont on its Nevada project—Labor shortages, new ethanol marketer deal." *Biofuels Digest*, October 28. http://www.biofuelsdigest.com/bdigest/2014/10/28/update-from-dupont-on-its-nevada-project-labor-shortages-new-ethanol-marketer-deal/.

Larsson, F., P. Andersson, and B.-E. Mellander. 2014. "Battery Aspects on Fires in Electrified Vehicles." Pp. 209-220 in *Proceedings from the Third International Conference on Fires in Vehicles—FIVE 2014, October 1-2, 2014, Berlin, Germany*. http://publications.lib.chalmers.se/records/fulltext/204919/local_204919.pdf.

Ley, J. 2016. "Unique Lanthanide-Free Motor Construction." UQM Technologies, Inc., Presentation to the 2016 U.S. Department of Energy Annual Merit Review, Washington, D.C., June 7. http://energy.gov/sites/prod/files/2016/06/f32/edt044_ley_2016_o_web.pdf.

Ludois, D. 2015. "Brushless and Permanent Magnet Free Wound Field Synchronous Motors for EV Traction." University of Wisconsin-Madison, Project EDT065, Presentation to the 2016 U.S. Department of Energy Annual Merit Review, Washington, D.C., June 9. http://energy.gov/sites/prod/files/2015/06/f24/edt065_ludois_2015_o.pdf.

Maness, P.C., and B. Logan. 2015. "Fermentation and Electro-hydrogenic Approaches to Hydrogen Production." National Renewable Energy Laboratory and Pennsylvania State University, Project PD038, Presentation to the 2015 U.S. Department of Energy Annual Merit Review, Washington, D.C., June 11.

Masiello, R. 2016. Transactive energy: The hot topic in the industry. *IEEE Power and Energy Magazine* 14(3):14-16. doi: 10.1109/MPE.2016.2525241.

Masiello, R., and J.R. Agüero. 2016. Sharing the ride of power: Understanding transactive energy in the ecosystem of energy economics. *IEEE Power and Energy Magazine* 14(3):70-78. doi: 10.1109/MPE.2016.2524965.

Masten, D., and D. Papageorgopoulos. 2016. "Fuel Cell Tech Team." Presentation to the Committee on the Review of the Research Program of the U.S. DRIVE Partnership, Phase 5, Washington, D.C., April 19.

Mayer, M. 2014. "From Prototype to Serial Production—Manufacturing Hydrogen Fueling Stations." Presentation to the A3PS Ecomobility 2014 Conference, Vienna, Austria, October 20. http://www.a3ps.at/a3ps-conferences.

McCormick, R., M. Ratcliff, E. Christensen, G. Chupka, J. Burton, T. Alleman, A. Marchese, D. Olsen, and M. Tian. 2015. "Performance of Biofuels and Biofuel Blends." National Renewable Energy Laboratory, Project FT003, Presentation at the 2015 U.S. Department of Energy Annual Merit Review, Washington, D.C., June 11. http://energy.gov/sites/prod/files/2015/06/f24/ft003_mccormick_2015_o.pdf.

McDaniel, A., and I. Ermanoski. 2015. "High Efficiency Solar Thermochemical Reactor for Hydrogen Production." Sandia National Laboratories, Project PD113, 2015 Annual Merit Review Proceedings: Hydrogen Production and Delivery, Washington, D.C., June 11. https://www.hydrogen.energy.gov/pdfs/review15/pd113_mcdaniel_2015_o.pdf.

Melaina, M.W., O. Antonia, and M. Penev. 2013. *Blending Hydrogen into Natural Gas Pipeline Networks: A Review of Key Issues*. Technical Report NREL/TP-5600-51995, March. http://www.nrel.gov/docs/fy13osti/51995.pdf.

Miller, E. 2016. "Hydrogen Production and Delivery Program." Plenary Presentation at the U.S. Department of Energy Annual Merit Review, Washington, D.C., June 6.

More, K. 2016. "New Fuel Cell Materials: Characterization and Method Development." Oak Ridge National Laboratory, Project FC020, Presentation at the 2016 U.S. Department of Energy Annual Merit Review, Washington, D.C., June 8.

Mueller, C. 2015. "Fuel Effects on Mixing-Controlled Combustion Strategies for High-Efficiency Clean-Combustion Engines." Combustion Research Facility: Sandia National Laboratories, Project FT004, Presentation at the 2015 U.S. Department of Energy Annual Merit Review, Washington, D.C., June 11. http://energy.gov/sites/prod/files/2015/06/f24/ft004_mueller_2015_o.pdf.

Myers, D. 2016. "Rationally Designed Catalyst Layers for PEMFC Performance Optimization." Argonne National Laboratory, Project FC106, Presentation at the 2016 U.S. Department of Energy Annual Merit Review, Washington, D.C., June 8.

National Biodiesel Board. 2016. "Production Statistics." http://biodiesel.org/production/production-statistics. Accessed October 26, 2016.

NRC (National Research Council). 2008. *Review of the Research Program of the FreedomCAR and Fuel Partnership, Second Report.* Washington, D.C.: The National Academies Press.

NRC. 2010. *Review of the Research Program of the FreedomCAR and Fuel Partnership, Third Report.* Washington, D.C.: The National Academies Press.

NRC. 2013. *Review of the Research Program of the U.S. DRIVE Partnership: Fourth Report.* Washington, D.C.: The National Academies Press.

NRC. 2014. *Reducing the Fuel Consumption and Greenhouse Gas Emissions of Medium and Heavy-Duty Vehicles—Phase Two: First Report.* Washington, D.C.: The National Academies Press.

NRC. 2015a. *Overcoming Barriers to Deployment of Plug-in Electric Vehicles.* Washington D.C.: The National Academies Press.

NRC. 2015b. *Review of the 21st Century Truck Partnership, Third Report.* Washington, D.C.: The National Academies Press.

NREL (National Renewable Energy Laboratory). 2017. *Co-Optimization of Fuels & Engines FY16 Year in Review.* http://www.nrel.gov/docs/fy17osti/67595.pdf.

Olejniczak, K.J. 2016. "Advanced Low-Cost SiC and GaN Wide Bandgap Inverters for Under-the-Hood Electric Vehicle Traction Drives." Cree Fayetteville, Inc., Project EDT058, Presentation to the 2016 U.S. Department of Energy Annual Merit Review, Washington, D.C., June 8. http://energy.gov/sites/prod/files/2016/06/f32/edt058_olejniczak_2016_o_web.pdf.

Omekanda, A. 2013. "Switched Reluctance Machines for EV and HEV Propulsion: State-of-the-Art." 2013 IEEE Workshop on Electrical Machines Design Control and Diagnosis (WEMDCD), Paris, France, March 11-12.

Peckham, J. 2014. Rivals to biofuels emerge. *Fuel*, June:36-40.

Peterson, D., and R. Farmer. 2016. "Historical Fuel Cell and Hydrogen Budgets." *DOE Hydrogen and Fuel Cells Program Record 16010*, March 16. https://www.hydrogen.energy.gov/pdfs/16010_historical_fuel_cell_h2_budgets.pdf.

Pitz, W., R. Whitesides, M. McNenly, M. Mehl, and N. Killingsworth. 2015. "Developing Kinetic Mechanisms for New Fuels and Biofuels, including CFD modeling." Lawrence Livermore National Laboratory, Project FT026, Presentation at the 2015 U.S. Department of Energy Annual Merit Review, Washington, D.C., June 11. http://energy.gov/sites/prod/files/2015/06/f24/ft026_pitz_2015_o.pdf.

POET. 2014. "First Commercial-Scale Cellulosic Ethanol Plant in the U.S. Opens for Business." http://www.poet.com/pr/first-commercial-scale-cellulosic-plant.

Ra, Y., P. Loeper, M. Andrie, R. Krieger, D. Foster, R. Reitz, and R. Durrett. 2012. Gasoline DICI engine operation in the LTC regime using triple-pulse injection. *SAE International Journal of Engines* 5(3):1109-1132.

Ramsden, T. 2015. "Pathway Analysis: Projected Cost, Lifecycle, Energy Use and Emissions of Emerging Hydrogen Technologies." National Renewable Energy Laboratory, Project SA036, Presentation to the 2015 U.S. Department of Energy Annual Merit Review, Washington, D.C., June 9.

Redon, F., C. Kalebjian, J. Kessler, N. Rakovec, J. Headley, G. Regner, and J. Koszewnik. 2014. "Meeting Stringent 2025 Emissions and Fuel Efficiency Regulations with an Opposed-Piston, Light-Duty Diesel Engine." SAE Technical Paper 2014-01-1187. doi:10.4271/2014-01-1187.

Ruth, M., and F. Joseck. 2011. "Hydrogen Threshold Cost Calculation." *Offices of Fuel Cell Technologies Program Record 11007*, March 25. https://www.hydrogen.energy.gov/pdfs/11007_h2_threshold_costs.pdf.

Sarkar, R. 2015. "Office of Energy Efficiency and Renewable Energy: FY16 Budget Overview." http://energy.gov/sites/prod/files/2015/03/f20/FY%202016%20Transportation%20Webinar%20%20Presentation.pdf.

Sarkar, R. 2016. "EERE Sustainable Transportation Overview." Presentation to the Committee on the Review of the Research Program of the U.S. DRIVE Partnership, Phase 5, Washington, D.C., February 3.

Satyapal, S. 2016. "U.S. Department of Energy Fuel Cell Technologies Office: Program Update for the National Research Council Review." Presentation to the Committee on the Review of the Research Program of the U.S. DRIVE Partnership, Phase 5, Washington, D.C., February 3.

Sjoberg, M. 2015. "Advanced Lean-burn, DI Spark Ignition Fuels Research." Combustion Research Facility: Sandia National Laboratories, Project FT006, Presentation at the 2015 U.S. Department of Energy Annual Merit Review, Washington, D.C., June 11. http://energy.gov/sites/prod/files/2015/06/f24/ft006_sjoberg_2015_o.pdf.

Slezak, L., and B. Gross. 2016. "Grid Interaction Technical Team." Presentation to the Committee on the Review of the Research Program of the U.S. DRIVE Partnership, Phase 5, Detroit, June 22.

Smith, B. 2012. *Chevrolet Volt Battery Incident Overview Report.* Report No. DOT HS 811573. Washington, D.C.: National Highway Traffic Safety Administration.

Solomon, A., and K. Howden. 2016. "Advanced Combustion and Emission Control Technical Team (ACECTT)." Presentation to the Committee on the Review of the Research Program of the U.S. DRIVE Partnership, Phase 5, Washington, D.C., April 19.

Stetson, N. 2014. "Hydrogen Storage Program Area: Plenary Presentation." U.S. Department of Energy, Fuel Cell Technologies Office, Presentation at the 2014 U.S. Department of Energy Annual Merit Review, Washington, D.C., June 16.

Stetson, N. 2015. "Hydrogen Storage Program Area: Plenary Presentation." U.S. Department of Energy, Fuel Cell Technologies Office, Presentation at the 2015 U.S. Department of Energy Annual Merit Review, Washington, D.C., June 8.

Stetson, N. 2016. "Hydrogen Storage Overview." U.S. Department of Energy, Fuel Cell Technologies Office, Presentation at the 2016 U.S. Department of Energy Annual Merit Review, Washington, D.C., June 6.

Stetson, N., and M. Veenstra. 2016. "Hydrogen Storage Tech Team Presentation to the NRC Committee on the Review of the U.S. DRIVE Partnership." Presentation to the Committee on the Review of the Research Program of the U.S. DRIVE Partnership, Phase 5, Washington, D.C., April 19.

Sutherland, E., and F. Joseck. 2015. "Low Volume Production and Delivery Cost." *DOE Hydrogen and Fuel Cells Program Record 15011*, August 20. https://www.hydrogen.energy.gov/pdfs/15011_low_volume_production_delivery_cost.pdf.

Szybist, J., S. Curran, S. Sluder, D. Splitter, A. Dempsey, and R. Wagner. 2015. "Gasoline-Like Fuel Effects on Advanced Combustion Regimes." Oak Ridge National Laboratory, Project FT008, Presentation at the 2015 U.S. Department of Energy Annual Merit Review, Washington, D.C., June 11. http://energy.gov/sites/prod/files/2015/06/f24/ft008_szybist_2015_o.pdf.

Taub, A.I., and A. Luo. 2015. Developing advanced lightweight materials and processes—Applying materials science to automotive engineering. *MRS Bulletin* 40(12):1045-1054.

Truett, R. 2015. GM, Honda cut cost of fuel cell stack. *Automotive News*, August 10. http://www.autonews.com/article/20150810000100/OEM01/308109956?template=print.

USCAR (United States Council for Automotive Research, LLC). 2014. News: Energy storage system goals. http://www.uscar.org/guest/article_view.php?articles_id=85.

U.S. DRIVE. 2013a. *Codes and Standards Technical Team Roadmap*. Washington, D.C.: U.S. Department of Energy. https://www1.eere.energy.gov/vehiclesandfuels/pdfs/program/cstt_roadmap_june2013.pdf.

U.S. DRIVE. 2013b. *Electrical and Electronics Technical Team Roadmap*. Washington, D.C.: U.S. Department of Energy. http://www1.eere.energy.gov/vehiclesandfuels/pdfs/program/eett_roadmap_june2013.pdf.

U.S. DRIVE. 2013c. *Electrochemical Energy Storage Technical Team Roadmap*. Washington, D.C.: U.S. Department of Energy. http://energy.gov/sites/prod/files/2014/02/f8/eestt_roadmap_june2013.pdf.

U.S. DRIVE. 2013d. *Fuel Pathway Integration Technical Team Roadmap*. Washington, D.C.: U.S Department of Energy. http://energy.gov/sites/prod/files/2014/02/f8/fpitt_roadmap_june2013.pdf.

U.S. DRIVE. 2013e. *Grid Interaction Technical Team Roadmap*. Washington, D.C.: U.S. Department of Energy. http://energy.gov/sites/prod/files/2014/02/f8/gitt_roadmap_june2013.pdf.

U.S. DRIVE. 2013f. *Hydrogen Delivery Technical Team Roadmap*. Washington, D.C.: U.S. Department of Energy. http://energy.gov/sites/prod/files/2014/02/f8/hdtt_roadmap_june2013.pdf.

U.S. DRIVE. 2013g. *Hydrogen Storage Technical Team Roadmap*. Washington, D.C.: U.S. Department of Energy. http://www1.eere.energy.gov/vehiclesandfuels/pdfs/program/hstt_roadmap_june2013.pdf.

U.S. DRIVE. 2013h. *USDRIVE Advanced Combustion and Emission Control Technical Team Roadmap*. Washington, D.C.: U.S. Department of Energy. http://www.energy.gov/sites/prod/files/2014/02/f8/acec_roadmap_june2013.pdf.

U.S. DRIVE. 2014. *Partnership Plan*. Washington, D.C.: U.S. Department of Energy. http://energy.gov/sites/prod/files/2014/12/f19/U.S.%20DRIVE%20Partnership%20Plan%20-%20December%202014%20with%20Addendum.pdf.

U.S. DRIVE. 2015. *Highlights of Technical Accomplishments*. Washington, D.C.: U.S. Department of Energy. http://energy.gov/sites/prod/files/2016/04/f30/2015%20U%20S%20%20DRIVE%20 Accomplishements%20Report.pdf.

Verbrugge, M.W., T.M. Lee, P.E. Krajewski, A.K. Sachdev, C. Bjelkengren, R. Roth, and R. Kirchain. 2009. Mass decompounding and vehicle lightweighting. *Materials Science Forum* 618-619: 411-418.

Weil, S., S. Dillich, F. Joseck, and M. Ruth. 2012. "H2 Production and Delivery Cost Apportionment." *DOE Hydrogen and Fuel Cells Program Record 12001*, May 14. https://www.hydrogen.energy.gov/pdfs/12001_h2_pd_cost_apportionment.pdf.

Weimer, A., B.J. Ward, B.D. Ehrhart, I. Al-Shankiti, C.L. Muhich, S.L. Miller, A.L. Hoskins, A. Groehn, and C.B. Musgrave. 2015. "Flowing Particle Bed Solarthermal Redox Process to Split Water." University of Colorado, Boulder, Project PD114, 2015 Annual Merit Review Proceedings: Hydrogen Production and Delivery, Washington, D.C., June 11. https://www.hydrogen.energy.gov/pdfs/review15/pd114_weimer_2015_o.pdf.

Williams, M. 2016. Honda's new motor for hybrid cars is a very big deal. *CIO-IDG News Service*, July 12. http://www.cio.com/article/3094891/car-tech/hondas-new-motor-for-hybrid-cars-is-a-very-big-deal.html.

Wimmer, R., and R. Gazelle. 2016. "Toyota Powertrain Strategy." Presentation to the Committee on the Review of the Research Program of the U.S. DRIVE Partnership, Phase 5, Washington, D.C., April 20.

Wu, F. 2016. "Materials Overview." U.S. Department of Energy Vehicle Technologies Office, Plenary Presentation to the 2016 U.S. Department of Energy Annual Merit Review, Washington, D.C., June 6.

Wu, G., A. Kolekar, K. Duho, D. Dotson, and F. Lockwood. 2015. "Improve Fuel Economy through Formulation Design and Modeling." Valvoline, Project FT025, Presentation at the 2015 U.S. Department of Energy Annual Merit Review, Washington, D.C., June 11. http://energy.gov/sites/prod/files/2015/06/f24/ft025_wu_2015_o.pdf.

Yamashita, A., M. Kondo, S. Goto, and O. Nobuyuki. 2015. "Development of High-Pressure Hydrogen Storage System for the Toyota 'Mirai.' " SAE Technical Paper 2015-01-1169, presented at the SAE 2015 World Congress and Exhibition, Detroit, April.

Yanowitz, J., M. Ratcliff, R. McCormick, J. Taylor, and M. Murphy. 2014. *Compendium of Experimental Cetane Numbers*. NREL Technical Report/TP-5400-61693. http://www.nrel.gov/docs/fy14osti/61693.pdf.

Zaluzec, M., and W. Joost. 2016. "Materials Technical Team." Presentation to the Committee on the Review of the Research Program of the U.S. DRIVE Partnership, Phase 5, Detroit, June 22.

Zhao, Z. 2016. "Next Generation Inverter." General Motors, Project EDT040, Presentation to the 2016 U.S. Department of Energy Annual Merit Review, Washington, D.C., June 7. http://energy.gov/sites/prod/files/2016/06/f32/edt040_zhao_2016_o_web.pdf.

Zhu, C. 2016. "High-Efficiency High-Density GaN-Based 6.6kW Bidirectional On-Board Charger for PEVs." DPM, Project EDT067, Presentation to the 2016 U.S. Department of Energy Annual Merit Review, Washington, D.C., June 8. http://energy.gov/sites/prod/files/2016/06/f32/edt067_zhu_2016_o_web.pdf.

Zibelman, A. 2016. REVing up the energy vision in New York: Seizing the opportunity to create a cleaner, more resilient, and affordable energy system. *IEEE Power and Energy* 14(3):18-24. doi: 10.1109/MPE.2016.2524967.

Zigler, B., J. Luecke, M. Ratcliff, G. Bogin, E. Osecky, J.Burton, P. Sindler, and D. Cameron. 2015. "Advanced Combustion and Fuels." National Renewable Energy Laboratory, Project FT002, Presentation at the 2015 U.S. Department of Energy Annual Merit Review, Washington, D.C., June 11. http://energy.gov/sites/prod/files/2015/06/f24/ft002_zigler_2015_o.pdf.

4

Overall Assessment

INTRODUCTION

The U.S. DRIVE Partnership has matured significantly since its creation. Since the National Research Council (NRC) Phase 4 report (2013), several improvements in Partnership operation have been made, some of them at least in part in reaction to prior NRC review recommendations. Noteworthy among these improvements are the addition of associate members to the technical teams, enhanced engagement by the Executive Steering Group (although still inadequate), and development of the cradle-to-grave (C2G) analysis capability.

In this final chapter in prior reviews, a detailed analysis has been made of the budget allocated to various initiatives and the budget trends over time. Since the Partnership has repeatedly stressed that in fact it has no budget and is simply one of many inputs to the Department of Energy (DOE) process of budgeting and managing its own projects, this level of detail seems to be less appropriate in this fifth review of the Partnership. Individual chapters have attempted to show the DOE funding allocated to various projects within the purview of the Partnership, but it must be stressed that these are DOE expenditures and not Partnership expenditures.

The task of forming an overall assessment is also complicated by the situation described in Chapter 2, whereby the Partnership and the DOE had difficulty defining for the committee just exactly what is included within the Partnership vis-à-vis the DOE portfolio of projects. Nevertheless, the committee has attempted in this Phase 5 report to review the appropriateness of, and progress on all of, the projects designated by the Partnership in which it is engaged, whether these precisely align with the DOE portfolio or not.

MAJOR ACHIEVEMENTS, PROGRESS, AND BARRIERS

Given that the Partnership exists primarily as a technical information exchange and serves as just one of the many inputs to DOE programs, where the budgets for all of this activity reside, it is difficult to ascertain exactly which achievements are directly attributable to the Partnership. The Partnership points to the DOE Annual Merit Review for details on all of its initiatives, where there is a wealth of valuable information.

Nevertheless, the individual chapters have documented achievements and barriers relating to their specific focus areas. Progress has clearly been made in such areas as advanced combustion, hydrogen fuel cell durability and cost, and electric drive systems (motor, power electronics, and battery) cost. At the same time, market introduction of improved hybrid electric vehicles and battery electric vehicles (BEVs) by automotive manufacturers represented in U.S. DRIVE and others, indicates that much of this technology is migrating out of the precompetitive realm and into the competitive marketplace. The hydrogen fuel cell vehicle (HFCV), currently being introduced in limited numbers by foreign original equipment manufacturers (OEMs), and expected by 2020 by one U.S. OEM (GM[1]), is expected to follow a path to commercialization with its own unique challenges, including, for example, infrastructure development. Since the Partnership is exclusively dedicated to precompetitive research and development, it is important that, informed by the C2G analysis, the portfolio be regularly reviewed to ensure that the focus remains on precompetitive challenges and relevant technology enablers. Some specific examples of areas where this critical review should take place are given in Chapter 3.

While some of the remaining challenges are purely technical, cost remains a formidable barrier for essentially all the technologies under development. The other notable barrier is the infrastructure challenge confronting hydrogen. Policy matters and deployment are by definition beyond the scope of the Partnership, but lack of infrastructure is arguably the biggest challenge to the widespread deployment of hydrogen fuel cell vehicles, and continued emphasis by DOE on infrastructure enablers as well as an implementation plan is vital, whether within the Partnership or not.

ADEQUACY, BALANCE, AND FUNDING

Historically, in the transition from the predecessor Partnership for a New Generation of Vehicles program into the FreedomCAR and Fuel Partnership, hydrogen-related activities represented roughly 70 percent of the relevant DOE budget, and non-hydrogen-related efforts consumed the remaining 30 percent. Coincident with the NRC (2010) Phase 3 report, a significant shift in this bal-

[1] See, for example, R. Truett, "Fuel Cell Puzzle Comes Together," *Automotive News*, October 11, 2016, http://www.autonews.com/article/20161011/BLOG06/310119999/fuel-cell-puzzle-comes-together.

ance took place and the hydrogen-related share dropped to roughly 30 percent. Having dropped by around 50 percent, the DOE budget devoted to hydrogen and fuel cells-related activities, which include both stationary and automotive applications, has since remained stable at around $100 million per year, while the budget devoted to non-hydrogen-related vehicle technology has gradually increased. For fiscal year 2016, hydrogen and fuel cell-related work is $101 million per year and the Vehicle Technologies Office funding is $310 million per year.

Much of the large shift in funding to vehicle technologies was in electric vehicles, especially batteries. Since fuel cell vehicles are inherently electric vehicles, much of the work on electric drivetrain and improved batteries is equally applicable to both plug-in electric vehicles (both BEVs and plug-in hybrid electric vehicles) and HFCVs, and while the $100 million devoted to purely hydrogen and fuel cell technologies is a much smaller share of the total DOE Energy Efficiency and Renewable Energy (EERE) budget, it is still felt by the committee to be appropriate as a share of the overall effort for projects supporting U.S. DRIVE targets and goals. Within that overall effort, priorities for funding may shift among technical areas as technical challenges change. Furthermore, as noted in Figure 4-1, there is hydrogen-related work in the DOE's Office of Basic Energy Sciences and Office of Fossil Energy, offices outside EERE, as well as the Advanced Research Projects Agency-Energy, which raised the overall total to approximately $150 million in FY 2015. (Similarly, there is work being done on batteries and electric drives within DOE but outside EERE.)

The respective DOE hydrogen/fuel cells and vehicle technologies budgets for fiscal year 2016 are shown in Figures 4-1 and 4-2, provided by DOE on February 3, 2016.

Finding 4-1. The share of Partnership-related DOE funding devoted to hydrogen/fuel cells has remained essentially stable at 25 to 30 percent since 2010 (having dropped from approximately 70 percent of total funding in earlier years) and is judged by the committee as appropriate, although within that overall effort, priorities for funding may shift among technical areas as technical challenges change.

CROSSCUTTING ISSUES

Since the Partnership is just one of many inputs to the DOE, and can offer only guidance and advice, the burden falls on the DOE leadership, assisted by the enhanced C2G analysis capability, to perform urgently needed in-depth portfolio analysis in two critical areas:

1. Review and, as appropriate, terminate those projects on technologies that no longer offer any realistic chance of meeting the objectives, and move related DOE funding to higher potential candidates. Several examples are cited throughout this report.

Key Activity	FY 15	FY 16	FY16
	($ in thousands)		
	Approp.	Request	Approp.
Fuel Cell R&D	33,000	36,000	35,000
Hydrogen Fuel R&D[1]	35,200	41,200	41,050
Manufacturing R&D	3,000	4,000	3,000
Systems Analysis	3,000	3,000	3,000
Technology Validation	11,000	7,000	7,000
Safety, Codes and Standards	7,000	7,000	7,000
Market Transformation	3,000	3,000	3,000
NREL Site-wide Facilities Support	1,800	1,800	1,900
Total	97,000	103,000	100,950

Office	FY 2015
EERE	$97.0M
Basic Science	$18.5M
Fossil Energy, SOFC	$30.0M

FY 2015 DOE Total: ~$150M

Number of Recipients funded from 2008-2015	
Industry	>110
Universities	>100
Laboratories	12

FIGURE 4-1 The Department of Energy's Energy Efficiency and Renewable Energy hydrogen and fuel cell budgets.
SOURCE: Satyapal (2016).

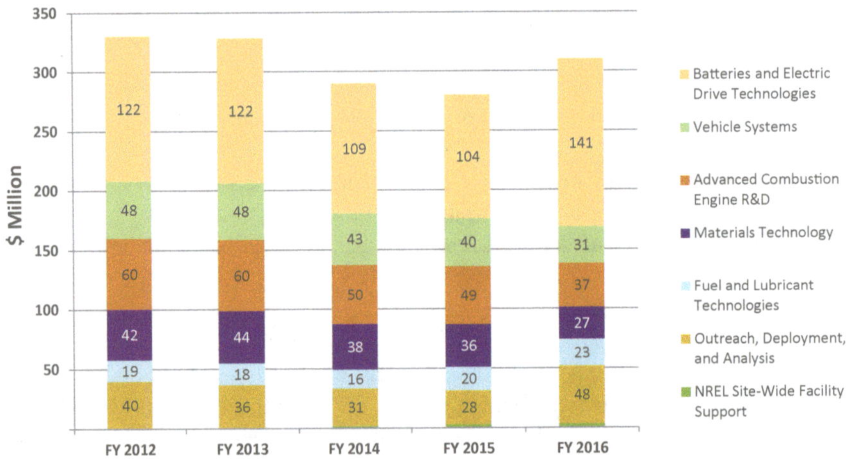

FIGURE 4-2 Department of Energy's Energy Efficiency and Renewable Energy Vehicle Technologies Office (VTO) budgets for fiscal year (FY) 2012 to 2016.
SOURCE: Howell (2016).

2. Drawing on advice from the technical teams, terminate activities that are crossing over into the competitive domain and shift the focus and funding to precompetitive technology enablers. While this analysis is urgent, it is anticipated that the resulting transition in activities will necessarily be gradual.

STRATEGIC ISSUES LOOKING FORWARD

Three trends, which could have strategic implications for the future, have emerged since the NRC Phase 4 review of the Partnership took place:

1. The dramatic change in domestic energy production has rendered the Obama administration's objective of reducing oil imports by 50 percent almost moot. While criteria pollutants will always be a concern and require substantial technical development to mitigate, it seems likely that in the future greenhouse gas (GHG) goals will present the greatest challenge. With GHG emissions as a primary focus, the pathways (e.g., combinations of vehicle technologies and fuels) to achieve the extremely aggressive goals are very limited and would suggest that Partnership-related projects be increasingly focused on those few pathways that offer a realistic chance of success in meeting those goals.
2. With numerous electric vehicles (HFCVs and BEVs) expected to enter the marketplace in the next few years, the consumer will be presented with a number of zero-emission vehicle options to select from. This transition can be expected to take many years, and infrastructure challenges are among the greatest challenges in each case, but particularly with regard to hydrogen. Although deployment and infrastructure are beyond the scope of the Partnership, there remains a need for precompetitive work on technology enablers to reduce system cost, improve durability, and substantially lower the cost of delivered "green" hydrogen at scale and electricity.
3. As discussed in Chapter 1 (see the section "Trends in Vehicle Automation and Smart Transportation"), the precise impact on the U.S. DRIVE Partnership is unclear at this point, but there is no doubt that the move toward connected and autonomous vehicles is dramatically accelerating. Somewhat related to this is the increasingly rapid proliferation of such personal mobility models as car-sharing and ride-sharing. While there does not appear to be an obvious connection between these trends and the current Partnership-related DOE portfolio, shared, autonomous, plug-in electric vehicles could contribute to the environmental and energy goals of U.S. DRIVE, and they deserve close scrutiny for their potential impact on the Partnership in the future.

Finding 4-2. Significant changes are occurring in the U.S. energy supply and demand situation, in the automotive sector as alternative propulsion system vehicles enter the marketplace, and in options for personal mobility, since the National Research Council Phase 4 review in 2012-2013.

Recommendation 4-1. The Executive Steering Group should identify appropriate changes in Partnership focus to reflect the impact of new personal mobility models, shrinking opportunities to achieve the aggressive greenhouse gas goals, the transition of many candidate technologies into the competitive domain, and the significant infrastructure challenges in providing hydrogen at fueling stations at a competitive cost—in particular, while retaining the focus on precompetitive technology enablers.

REFERENCES

Howell, D. 2016. "DOE/EERE Vehicle Technologies Office Overview." Presentation to the Committee on the Review of the Research Program of the U.S. DRIVE Partnership, Phase 5, Washington, D.C., February 3.

NRC (National Research Council). 2010. *Review of the Research Program of the FreedomCAR and Fuel Partnership, Third Report*. Washington, D.C.: The National Academies Press.

NRC. 2013. *Review of the Research Program of the U.S. DRIVE Partnership, Fourth Report*. Washington, D.C.: The National Academies Press.

Satyapal, S. 2016. "Program Update for the National Research Council Review." Presentation to the Committee on the Review of the Research Program of the U.S. DRIVE Partnership, Phase 5, Washington, D.C., February 3.

Apppendixes

A

Biographic Sketches of Committee Members

John H. Johnson, *Chair*, is a presidential professor emeritus in the Department of Mechanical Engineering-Engineering Mechanics at Michigan Technological University (MTU) and fellow of the Society of Automotive Engineers (SAE) and the American Society of Mechanical Engineers (ASME). His experience spans a wide range of analysis and experimental work related to advanced engine concepts, diesel and other internal engine emissions studies, fuel systems, and engine simulation. He was previously project engineer at the U.S. Army Tank Automotive Center and chief engineer in applied engine research at the International Harvester Company before joining the MTU mechanical engineering faculty. He served as chairman of the MTU mechanical engineering and engineering mechanics department from 1986 to 1993. He has served on many committees related to engine technology, engine emissions, and health effects—for example, committees of the SAE, the National Academies of Sciences, Engineering, and Medicine, the Combustion Institute, the Health Effects Institute, and the Environmental Protection Agency—and consults to a number of government and private-sector institutions. In particular, he served on many National Academies committees, including the Committee on Fuel Economy of Automobiles and Light Trucks, the Committee on Advanced Automotive Technologies Plan, the Committee on the Impact and Effectiveness of Corporate Average Fuel Economy (CAFE) Standards, and the Committee to Assess Fuel Economy for Medium and Heavy-Duty Vehicles. He has served as chair of several committees, most recently for the Committee on Review of the 21st Century Truck Partnership, Phase 3. Dr. Johnson received from SAE the Horning Memorial Award, Colwell Merit Award (two), McFarland Award, Myers Award for Outstanding Student Paper, the Franz Pischinger Powertrain Innovation Award, and from ASME the Honda Medal and the Internal Combustion Engine Award. He received his Ph.D. in mechanical engineering from the University of Wisconsin.

Alexis T. Bell is a Dow Professor of Sustainable Engineering at the University of California, Berkeley. He has also held the positions of dean, College of Chemistry, and chairman, Department of Chemical Engineering. The emphasis of his research is on heterogeneous catalysis and the relationship between catalyst composition and structure and catalyst performance on the molecular level. He has developed and applied experimental and theoretical methods to understanding catalysts and catalyzed reactions at the molecular level. His insights can be used to improve catalyst activity and selectivity for applications such as air-pollution control and the synthesis of fuels and chemicals. He is a recipient of the Curtis W. McGraw Award for Research, American Association of Engineering Education; the Professional Progress, R. H. Wilhelm, and William H. Walker Awards, American Institute of Chemical Engineers; the Paul H. Emmett Award in Fundamental Catalysis and the Michel Boudart Award, Catalysis Society; and the American Chemical Society (ACS) Award for Creative Research in Homogeneous or Heterogeneous Catalysis and the George Olah Award in Petroleum or Hydrocarbon Chemistry, ACS. He has served on a number of National Academies' boards and committees, and is a member of the National Academy of Engineering (NAE), a member of the National Academy of Sciences, and a fellow of the American Association for the Advancement of Science. He received his Sc.D. in chemical engineering from the Massachusetts Institute of Technology (MIT).

David L. Bodde serves as a professor and senior fellow at the International Center for Automotive Research at Clemson University. Prior to joining Clemson University, Dr. Bodde held the Charles N. Kimball Chair in Technology and Innovation at the University of Missouri in Kansas City. Dr. Bodde serves on the board of directors of several energy and technology companies, including Great Plains Energy and the Commerce Funds. His executive experience includes vice president, Midwest Research Institute; assistant director of the Congressional Budget Office; and deputy assistant secretary in the U.S. Department of Energy (DOE). He has served as a member of a number of National Academies' boards and committees, including the Board on Energy and Environmental Systems, the Committee on Alternatives and Strategies for Future Hydrogen Production and Use, the Committee on Assessment of Resource Needs for Fuel Cell and Hydrogen Technologies, and the Committee on Review of the U.S. DRIVE Research Program, Phase 4. He has a doctorate in business administration from Harvard University, M.S. degrees in nuclear engineering (1972) and management (1973), and a B.S. from the United States Military Academy. Upon graduation, he served in the U.S. Army in Vietnam.

Nady Boules is president of NB Motors, LLC, an engineering and management consulting firm. Prior to this, he held several positions in the automotive industry including director, Electrical and Control Systems Research Laboratory, General Motor Global R&D; co-director, GM-Carnegie Mellon University VIT Collaborative Research Laboratory; director, Innovation and Technology Leadership, and director, Dynamics Innovation Center and Materials Engineering, Delphi Corporation; section manager, Electromechanical Systems, and group leader, Electric Machines and Actuators, General Motors Global R&D Operations; and senior magnetics engineer, Simmonds Precision Corporation, Engine Systems Division. He has expertise in a number of areas, including vehicle electrification, electric drives and electric propulsion systems, and electrical and electromechanical systems and subsystems. He has been on the board of directors, Intelligent Transportation Systems of America, the board of directors, Electricore, Inc., and the board of advisors, Institute for Advanced Vehicle Systems, University of Michigan. He has been awarded the Institute of Electrical and Electronics Engineers Nikola Tesla Award (2011), the Delphi Automotive Systems PACE Award (1998), and the GM President's Council Honors Award (1996). He has a B.S.E.E. and an M.S.E.E. from Cairo University, and a Ph.D. (Dr.-Ing.) in electrical engineering from the University of Braunschweig, Germany.

Glenn A. Eisman is the principal partner at Eisman Technology Consultants, LLC; adjunct professor of materials science, Rensselaer Polytechnic Institute; and adjunct professor, Department of Engineering, Union Graduate College. His previous positions include chief executive officer and chief technology officer, H2Pump LLC; chief technology officer, Plug Power, Inc.; technical leader, Advanced Materials Program, Dow Chemical Company; project leader, Discovery Research R&D and Inorganic Chemicals Research, Dow Chemical Company; and a Robert A. Welch research fellow, University of Texas, Austin. Dr. Eisman has over 35 years of experience in R&D and product development in fuel cells, hydrogen technologies, electrochemical engineering, physical and inorganic solid state chemistry, and new technology commercialization and business development. He received the Inventor of the Year Award from Dow Chemical Company in 1993. He received a B.S. in chemistry, Temple University, and a Ph.D. in physical inorganic chemistry, Northeastern University. He holds 22 patents and has published 42 papers.

David E. Foster, the Phil and Jean Myers Professor Emeritus of Mechanical Engineering, received his B.S. and M.S. degrees in mechanical engineering from the University of Wisconsin, Madison, in 1973 and 1975, respectively. He received his Ph.D. in mechanical engineering in 1979 from MIT. He was a faculty member at the University of Wisconsin from completion of his Ph.D. until he retired in 2012. He is an active member of the Engine Research Center, of which he served as the director from 1994 through 1999, and from September 2008 through December

2011. He was also the founding co-director of the General Motors–ERC Collaborative Research Laboratory, from its inception in 2002 until his retirement. Dr. Foster is a registered professional engineer in the state of Wisconsin and has won departmental, engineering society, and university awards for his classroom teaching. He was a member of the National Academies PNGV Review Committee for 6 years and has served on the Committee to Assess Fuel Economy Technologies of Medium and Heavy-Duty Vehicles, the Committee to Review the DOE FreedomCAR and Fuels Partnership Program, the 21st Century Truck Review, and the U.S. DRIVE Program Review. He has been the recipient of the Academic Contribution Award from the Japan Society of Automotive Engineers, the UW Engineering Byron Bird Excellence in Research Publication Award, the ASME Honda Gold Medal for outstanding contributions in the field of personal transportation, and the 2011 SAE Horning Award, and he is a fellow of SAE.

Matt Fronk is president of Matt Fronk & Associates, LLC. He has more than 37 years of experience leading both research and product development projects in advanced technology, fuel cells, and energy storage. He spent 20 years leading General Motors' Fuel Cell Research and Development program in Honeoye Falls, New York (Monroe County). During his tenure at GM, fuel cell systems were developed from laboratory-scale systems to 100 operating vehicles—the largest of any original equipment manufacturer auto company at the time. Mr. Fronk also has extensive global supplier development experience. After GM, he served as director of the Center for Sustainable Mobility at Rochester Institute of Technology and was instrumental in developing durability and life cycle analyses for new product designs as they moved from concept to product. He also was a founding member and first board chair for NY BEST—an energy storage consortium in New York State—and continues to this day as a board member. He led the design and build of the NY BEST Battery Test Center in Rochester, New York, a state-of-the-art facility that opened in April 2014. He served recently on the National Academies Committee on Fuel Economy Technologies for Light-Duty Vehicles, Phase 2. Mr. Fronk is an expert consultant to the energy storage and fuel cell fields and co-chairs the Energy Innovation Economic development group in the Finger Lakes region of New York. He has a B.S. in mechanical engineering from Union College.

Robert J. Nowak is a consultant and former program manager at the Defense Advanced Research Projects Agency and the Office of Naval Research and a staff scientist and section head at the Naval Research Laboratory. He has directed and supported research in fuel cells, batteries, capacitors, energy harvesting, fuel processing, thermal energy conversion, micro-engines, hydrogen storage, biofuel cells, sonoluminescence, and biomolecular motors. He has served on eight National Academies committees, which used his expertise in energy-related topics. His experience is in fuel cells, fuel processing, batteries, and hydrogen production and storage. He received his B.A. and M.S. degrees in chemistry from

Oakland University and his Ph.D. degree in chemistry from the University of Cincinnati. He performed postdoctoral work at the University of North Carolina, Chapel Hill, and the Naval Research Laboratory, where he was selected through the National Research Council Postdoctoral Program. He received the Department of Defense Meritorious Civilian Service Award for his efforts to develop portable power options for the military.

Bernard Robertson is retired from DaimlerChrysler Corporation. During the latter part of his 38-year career in the automotive industry, he was elected an officer of Chrysler Corporation in February 1992. He was appointed senior vice president coincident with the merger of Chrysler Corporation and Daimler-Benz AG in 1998, and was named senior vice president of Engineering Technologies and Regulatory Affairs in 2001. In his last position he led the Liberty and Technical Affairs Research Group; Advanced Technology Management and FreedomCAR activities; and hybrid electric, battery electric, fuel cell, and military vehicle development. In addition, he was responsible for regulatory analysis and compliance for safety and emissions. He has served on a number of National Academies' committees, most recently on the Committee on Review of the 21st Century Truck Partnership, Phase 3, and the Committee on Review of the U.S. DRIVE Research Program, Phase 4. Mr. Robertson holds an M.B.A. degree from Michigan State University, a master's degree in automotive engineering from the Chrysler Institute, and a master's degree in mechanical sciences from Cambridge University, England. He is a member of the NAE, a fellow of the Institute of Mechanical Engineers (UK), a chartered engineer (UK), and a fellow of the SAE.

James A. (Jim) Spearot is currently president of his own consulting company, Mountain Ridgeline Consulting, LLC. His consulting efforts focus on transportation energy and automotive fuel and lubricant issues as they affect emissions and fuel efficiency. In 2009, Dr. Spearot retired from General Motors Research and Development Center, where he was director of the Chemical and Environmental Sciences Laboratory, whose mission was to develop cost-effective environmental strategies and systems for GM's products and processes. Additionally, Dr. Spearot served as chief scientist for GM's Public Policy Center, lead executive for research programs in Russia and Commonwealth of Independent States countries, and manager of GM's Hydrogen Storage Innovation Program. Dr. Spearot began his GM career in 1972 as an assistant senior research engineer in the Fuels and Lubricants Department. He was appointed department head of Fuels and Lubricants in 1992 and director of the Chemical and Environmental Sciences Laboratory in 1998. He is a member of several organizations: SAE, the Society of Rheology, the Society of Tribologists and Lubrication Engineers, and the American Institute of Chemical Engineers. He is a former chairman of the SAE Fuels and Lubricants Division and a former chairman of the Coordinating Research Council. He has served as chairman of the Fuels Working Group of the

U.S. Council for Automotive Research (USCAR) and the USCAR Environmental and Hydrogen Technical Leadership Councils. His professional honors include an ASTM Award for Excellence in 1990; the Arch T. Colwell Merit Award from SAE in 1987; and the Award for Research on Automotive Lubricants, also from the SAE, in 1987. He is a fellow member of the SAE and has received a Lifetime Achievement Award from USCAR. He recently served on the National Academies Committee on Review of the 21st Century Truck Partnership, Phase 3. He holds a B.S. in chemical engineering from Syracuse University and master's and doctorate degrees, also in chemical engineering, from the University of Delaware.

Satish Tamhankar retired as a technology expert from Linde, LLC, Technology and Innovation. Prior to joining Linde in 2007, he held several positions at BOC Gases, Process Gas Solutions Technology, including section director and technology manager, technology fellow, principal scientist, and senior lead scientist. Prior to working at BOC Gases, he was also a research faculty member at the California Institute of Technology. His expertise covers a wide range of energy technology areas, including waste and biomass conversion to energy and fuels; energy storage; hydrogen production, storage, dispensing and fuel cell applications; selective oxidation processes; and a variety of chemical process improvement technologies. He has extensive expertise in applied catalysis and chemical reaction engineering; extensive R&D experience from concept to pilot scale, including commercial and business development support; and holds more than 30 U.S. patents and has more than 40 technical publications. He received the 2007 Patent Award from the Linde Group Inventors Club; the 2003 BOC Innovation Achievement Award for catalytic partial oxidation process development; and the 1994 Airco Technology Innovation Award for inert gas purifier development. He has a B.Sc., chemistry, and an M.Sc. and Ph.D., physical chemistry, from Pune University, India. He conducted postdoctoral research in chemical engineering at West Virginia University, Morgantown.

Alan Taub is professor of materials science and engineering, College of Engineering, University of Michigan, and chief technology officer of LIFT (Lightweight Innovations for Tomorrow). Formerly, he was vice president, global R&D, General Motors Company. He has focused on automotive technology innovation. Areas of particular interest have been lightweight materials for structural applications, active and passive safety for vehicles, advanced propulsion systems, and simulation of vehicle performance. Within lightweight materials, emphasis has been on processing and joining of high-strength steel, aluminum, and magnesium alloys. Activities in vehicle safety have concentrated on mechanical properties of vehicle body structure, especially high strain rate deformation and, more recently, sensors and controls for accident avoidance. He has been broadly involved in propulsion systems ranging from improving efficiency of internal combustion engines to proton exchange membrane fuel cells and hydrogen storage. He has a strong technical interest in computer-aided engineering tools to simulate structural performance. Earlier technical work in his career centered on interaction of microstructure and mechanical deformation with electrical properties of amorphous alloys and superconductors. He was elected to the NAE for contributions to the development of innovative electrical materials and automotive technologies and leadership in the globalization of automotive research. He has a B.Sc. in materials engineering, Brown University, and an M.S. and a Ph.D. in physics from Harvard University.

Kathleen C. Taylor is retired from the General Motors Research Laboratories in Warren, Michigan, where she worked for 31 years. Her last assignment was director of the Materials and Processes Laboratory and simultaneously chief scientist for General Motors of Canada, Ltd., in Oshawa Ontario. Earlier she was department head for Physics and Physical Chemistry and department head for Environmental Sciences. She recently served on the DOE Hydrogen Technology Advisory Committee and the NRC review of the 21st Century Truck Partnership. She has expertise in research and development management, fuel cells, batteries, catalysis, exhaust emission control, and automotive materials. Dr. Taylor was awarded the Garvan Medal from the American Chemical Society. She is a member of the NAE and a fellow of SAE, the American Academy of Arts and Sciences, and the Indian National Academy of Engineering. She was president of the Materials Research Society and chair of the board of directors of the Gordon Research Conferences. She received an A.B. in chemistry from Douglass College and a Ph.D. in physical chemistry from Northwestern University.

Brijesh Vyas retired as a distinguished member of the technical staff at Bell Laboratories, research division of AT&T, Lucent Technologies, Alcatel-Lucent and LGS Innovations, LLC. He was the technical manager of the energy conversion technology group responsible for research and development of advanced materials and technologies for energy storage systems. He has led efforts to develop various rechargeable batteries and related energy conversion technologies for a variety of telecommunications applications. He was formerly at the Brookhaven National Laboratory and has been a guest professor at the Technical University of Denmark in Copenhagen investigating corrosion and erosion of metals. He received the Sam Tour Award from the American Society of Materials and Testing. His areas of expertise include materials science, electrochemistry, energy storage, and corrosion. He served on the National Academies Committee to Review the U.S. Advanced Battery Consortium's Electric Vehicle Battery R&D Project Selection process, and the Committee on Review of the U.S. DRIVE Research Program, Phase 4. He received a bachelor's degree in metallurgical engineering from the Indian Institute of Technology in Bombay and a Ph.D. in materials science from the State University of New York, Stony Brook.

B

U.S. Department of Energy, Energy Efficiency and Renewable Energy Organization Chart (as of September 2016)

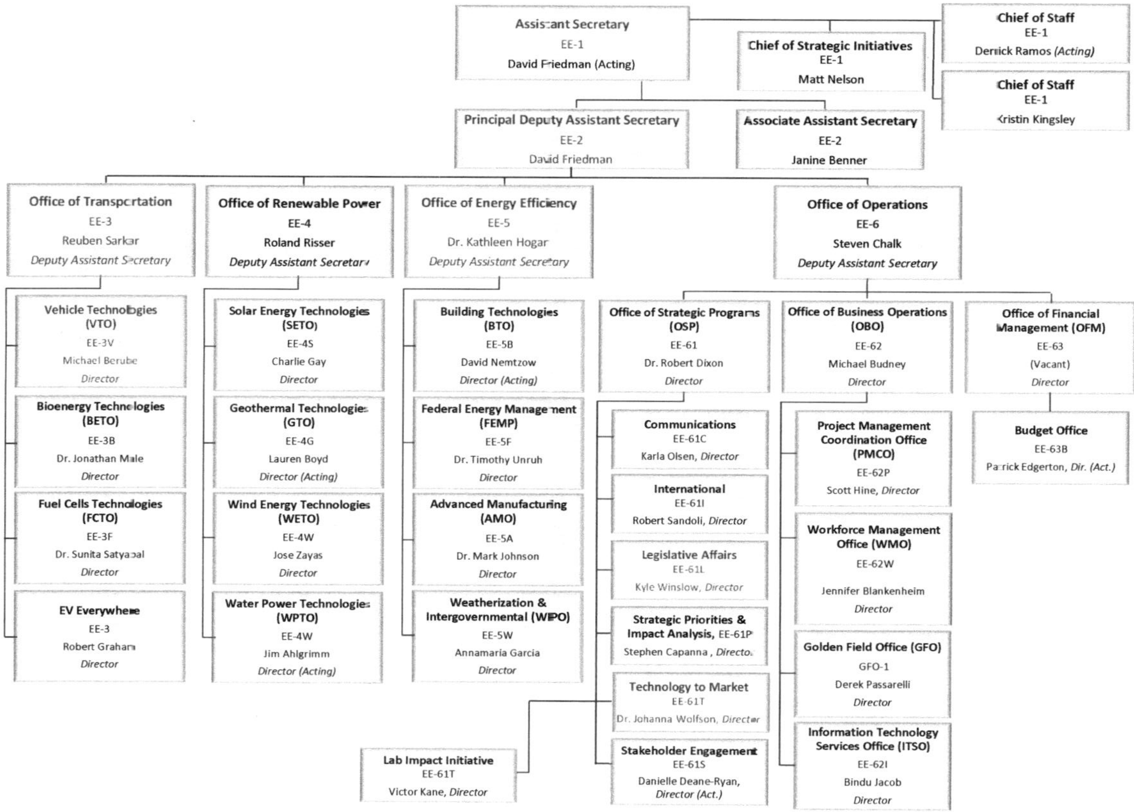

FIGURE B-1 Organizational chart of the U.S. Department of Energy's Office of Energy Efficiency and Renewable Energy.

C

Meetings and Presentations

FIRST COMMITTEE MEETING
FEBRUARY 3-4, 2016

U.S. DRIVE Overview Presentation
Christy Cooper, DOE Director, U.S. DRIVE Partnership, and Acting Director, Vehicle Technologies Office (VTO)

Overview of the Office of Transportation in the Office of Energy Efficiency and Renewable Energy (EERE)
Reuben Sarkar, Deputy Assistant Secretary for Transportation, EERE

Overview of the Vehicle Technologies Office (VTO)
David Howell, Program Manager for Hybrid Electric Systems

Overview of the Fuel Cell Technologies Office (FCTO)
Sunita Satyapal, Director, FCTO

Vehicle Operations Group Perspective on U.S. DRIVE
Steve Zimmer, Executive Director, USCAR

Utility Operations Group Perspective on U.S. DRIVE
Dan Bowermaster, Electric Power Research Institute

Energy Company Perspective on U.S. DRIVE
Jim Simnick, Technical Advisor, BP Global Fuels Technology

SECOND COMMITTEE MEETING
APRIL 19-20, 2016

Target Setting Process
Jacob Ward, Lead Analyst, DOE/EERE Vehicle Technologies Office (VTO)

Vehicle Systems Analysis Technical Team (VSATT)
David Anderson, VSATT Co-chair
Norman Bucknor, VSATT Industry Co-Chair (General Motors)

Fuel Pathway Integration Technical Team (FPITT)
Fred Joseck, FPITT DOE Co-chair
Laura Verduzco, FPITT Industry Co-Chair (Chevron)

Hydrogen Delivery Technical Team (HDTT)
Erika Gupta, HDTT DOE Co-chair
Herie Soto, HDTT Industry Co-chair (Shell)

Hydrogen Production Technical Team (HPTT)
Katie Randolph, HPTT DOE Co-chair
Bryan Chapman, HPTT Industry Co-chair (ExxonMobil)

Hydrogen Storage Technical Team (HSTT)
Ned Stetson, HSTT DOE Co-chair
Mike Veenstra, HSTT Industry Co-chair (Ford Motor Company)

Fuel Cells Technical Team (FCTT)
Dimitrios Papageorgopoulos, FCTT DOE Co-chair
David Masten, FCTT Industry Co-chair (General Motors)

Electrochemical Energy Storage Tech Team (EESTT)
David Howell, EESTT DOE Co-chair
Kent Snyder, EESTT Industry Co-chair (Ford)

Advanced Combustion and Emissions Control Tech Team (ACECTT)
Ken Howden, ACECTT DOE Co-chair
Arun Solomon, ACECTT Industry Co-chair (General Motors)

Toyota's Powertrain Strategy
Robert Wimmer, Toyota
Rick Gezelle, Toyota

THIRD COMMITTEE MEETING
JUNE 22-23, 2016

Electrical/Electronics Technical Team (EETT)
Robert Dawsey, EETT Industry Co-chair (GM)
Susan Rogers, EETT DOE Co-chair

Grid Interaction Technical Team (GITT)
Lee Slezak, GITT DOE Co-chair

Fuels Working Group (FWG)
Jeff Farenback-Brateman, FWG Industry Co-chair (ExxonMobil)
Kevin Stork, FWG DOE Co-chair
William Studzinski, FWG Industry Co-chair (GM)

Materials Technical Team (MTT)
William Joost, MTT DOE Co-chair
Matthew Zaluzec, MTT Industry Co-chair (Ford)

Codes and Standards Technical Team (CSTT)
Will James, CSTT DOE Co-chair
Ian Sutherland, CSTT Industry Co-chair (GM)

Cradle-to-Grave Analysis Working Group (C2G)
Fred Joseck, C2G DOE Co-chair
Tim Wallington, Industry Co-chair (Ford)
Jacob Ward, C2G DOE Co-chair

Hydrogen Production and Distribution
David Edwards, Director of Technology Partnership, Air Liquide

FOURTH COMMITTEE MEETING
OCTOBER 6-7, 2016

No open session presentations.

D

Acronyms

AC	alternating current
ACECTT	advanced combustion and emission control technical team
AMR	Annual Merit Review
ANL	Argonne National Laboratory
APEEM	Advanced Power Electronics and Electric Motors
ARPA-E	Advanced Research Projects Agency-Energy
ASCR	Advanced Scientific Computing Research
BatPaC	Battery Performance and Cost Model
BATT	Batteries for Advanced Transportation Technologies
BES	Office of Basic Energy Sciences
BETO	Bioenergy Technologies Office
BEV	battery electric vehicle
BFE	brushless field excited
BLDC	brushless permanent magnet DC motor
BMEP	brake mean effective pressure
BNL	Brookhaven National Laboratory
BTE	brake thermal efficiency
CAMP	Cell Analysis, Modeling, and Prototyping
C2G	cradle to grave
CAEBAT	Computer-Aided Engineering for Electric Drive Vehicle Batteries
CAFE	Corporate Average Fuel Economy
CAS	complex adaptive system
CAV	connected and autonomous vehicle

CCM	comparative cost metric
CCSU	carbon capture, storage and utilization
CERC	Clean Energy Research Center
CFD	computational fluid dynamics
CI	compression ignition
CLEERS	Cross Cut Lean Exhaust Emissions Reduction Simulations
CNG	compressed natural gas
CO	carbon monoxide
CO_2	carbon dioxide
CRC	Coordinating Research Council
CSTT	codes and standards technical team
DC	direct current
DCN	derived cetane number
DER	distributed energy resources
DOE	Department of Energy
DOT	Department of Transportation
DRIVE	Driving Research and Innovation for Vehicle Efficiency and Energy Sustainability
EEN	emergent entrepreneurial network
EERE	Energy Efficiency and Renewable Energy
EES	electrochemical energy storage
EESTT	electrochemical energy storage technical team
EGR	exhaust gas recirculation
EHC	electrochemical hydrogen compression
EIA	Energy Information Administration
EIT	engineered ionomer topology
EPA	Environmental Protection Agency
EPRI	Electric Power Research Institute
ESG	Executive Steering Group
EVI	Electric Vehicles Initiative
FACA	Federal Advisory Committee Act
FAME	fatty acid methyl ester
FCE	fuel cell energy
FCEV	fuel cell electric vehicle
FC-PAD	Fuel Cell Consortium for Performance and Durability
FCTO	Fuel Cell Technologies Office
FCTT	fuel cell technical team
FEERC	Fuels, Engines and Emissions Research Center
FPITT	fuel pathway integration technical team
FWG	Fuels Working Group

FY	fiscal year
g	gram
GDI	gasoline direct injection
GGE	gasoline gallon equivalent
GHG	greenhouse gas
GITT	grid interaction technical team
GREET	Greenhouse Gases, Regulated Emissions, and Energy Use in Transportation
GTDI	gasoline turbocharged direct injection
GTR	Global Technical Regulation
H or H_2	hydrogen
H2A	Hydrogen Analysis
HC	hydrocarbon
HDSAM	Hydrogen Delivery Scenario Analysis Model
HEV	hybrid electric vehicle
HFCV	hydrogen fuel cell vehicle
HPTT	hydrogen production technical team
HSECoE	Hydrogen Storage Engineering Center of Excellence
HT	Hydrogenious Technologies
HyMARC	Hydrogen Materials—Advanced Research Consortium
IAPG	Interagency Advanced Power Group
ICE	internal combustion engine
IEA	International Energy Agency
IM	induction motor
IQT	Ignition Quality Tester
IR	infrared
JOG	Joint Operations Group
kg	kilogram
kW	kilowatt
kWh	kilowatt-hour
LANL	Los Alamos National Laboratory
LBNL	Lawrence Berkeley National Laboratory
LH2	liquid hydrogen
Li ion	lithium ion
LLNL	Lawrence Livermore National Laboratory
LTC	low-temperature combustion
LTGC	low-temperature gasoline combustion

MEA	membrane electrode assembly
MHC	metal-hydride compression
MOF	metal-organic framework
MOU	memorandum of understanding
mpg	miles per gallon
MTT	materials technical team
NAE	National Academy of Engineering
NAS	National Academy of Sciences
NETL	National Energy Technology Laboratory
NFPA	National Fire Protection Association
NHTSA	National Highway Traffic Safety Administration
NIST	National Institute of Science and Technology
NO	nitrogen oxide
NO_x	nitrogen oxides
NRC	National Research Council
NREL	National Renewable Energy Laboratory
NVH	noise, vibration, and harshness
OEM	original equipment manufacturer
ORNL	Oak Ridge National Laboratory
PAN	polyacrylonitrile
PB	photobiological
PEC	photoelectrochemical
PEIS	Programmatic Environmental Impact Statement
PEMFC	proton exchange membrane fuel cell
PGM	platinum group metal
PHEV	plug-in hybrid electric vehicle
PNGV	Partnership for a New Generation of Vehicles
PNNL	Pacific Northwest National Laboratory
PV	photovoltaic
QTR	Quadrennial Technology Reviews
R&D	research and development
RCCI	reactivity controlled combustion ignition
RCM	rapid compression machine
RE	rare earth
RFS	Renewable Fuel Standard
RON	research octane number
SA	Strategic Analysis, Inc.

SBIR	Small Business Innovation Research
SCR	selective catalytic reduction
SCS	Safety, Codes and Standards
SEI	solid-electrolyte interphase
SGDI	spray-guided gasoline direct-injection
SI	spark ignition
SiC	silicon carbide
SMR	steam methane reforming
SNL	Sandia National Laboratories
SRM	switched reluctance motor
STCH	solar thermochemical hydrogen
SUV	sport utility vehicle
SyRM	synchronous reluctance motor
21CTP	21st Century Truck Partnership
TDS	total drive system/traction drive system
TE	transactive energy
TSTF	target setting task force
USABC	U.S. Advanced Battery Consortium
USCAR	U.S. Council for Automotive Research
UV	ultraviolet
V2G	vehicle-to-grid
VCR	variable compression ratio
VSATT	vehicle systems analysis technical team
VTO	Vehicle Technologies Office
WBG	wide bandgap
XANES	X-ray absorption near-edge structure
ZEV	zero-emission vehicle